INVESTIGATIONS IN NUMBER, DATA, AND SPACE®

Addition and Subtraction

Number Games and Story Problems

Grade 1

Also appropriate for Grade 2

Marlene Kliman
Susan Jo Russell

Developed at TERC, Cambridge, Massachusetts

Dale Seymour Publications®
Menlo Park, California

The *Investigations* curriculum was developed at TERC (formerly Technical Education Research Centers) in collaboration with Kent State University and the State University of New York at Buffalo. The work was supported in part by National Science Foundation Grant No. ESI-9050210. TERC is a nonprofit company working to improve mathematics and science education. TERC is located at 2067 Massachusetts Avenue, Cambridge, MA 02140.

This project was supported, in part,
by the
National Science Foundation
Opinions expressed are those of the authors
and not necessarily those of the Foundation

Managing Editor: Catherine Anderson
Series Editor: Beverly Cory
ESL Consultant: Nancy Sokol Green
Production/Manufacturing Director: Janet Yearian
Production/Manufacturing Manager: Karen Edmonds
Production/Manufacturing Coordinator: Amy Changar
Design Manager: Jeff Kelly
Design: Don Taka
Composition: Andrea Reider, Thomas Dvorak
Illustrations: DJ Simison, Carl Yoshihara, Rachel Gage
Cover: Bay Graphics

This book is published by Dale Seymour Publications®, an imprint of Addison Wesley Longman, Inc.

Dale Seymour Publications
2725 Sand Hill Road
Menlo Park, CA 941025
Customer Service: 800-872-1100

Order number DS43707
ISBN 1-57232-470-8
1 2 3 4 5 6 7 8 9 10-ML-01 00 99 98 97

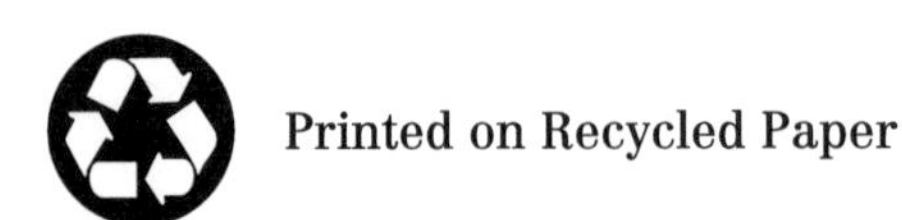

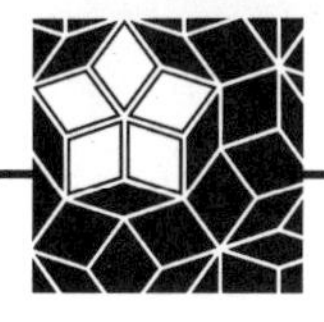

INVESTIGATIONS IN NUMBER, DATA, AND SPACE®

TERC

Principal Investigator **Susan Jo Russell**
Co-Principal Investigator **Cornelia C. Tierney**
Director of Research and Evaluation **Jan Mokros**
Director of K–2 Curriculum **Karen Economopoulos**

Curriculum Development
Karen Economopoulos
Marlene Kliman
Jan Mokros
Megan Murray
Susan Jo Russell
Tracey Wright

Evaluation and Assessment
Mary Berle-Carman
Jan Mokros
Andee Rubin

Teacher Support
Irene Baker
Megan Murray
Judy Storeygard
Tracey Wright

Technology Development
Michael T. Battista
Douglas H. Clements
Julie Sarama

Video Production
David A. Smith
Judy Storeygard

Administration and Production
Irene Baker
Amy Catlin

Cooperating Classrooms for This Unit
Malia Scott
Brookline Public Schools
Brookline, MA

Maryellen Bertrand
Boston Public Schools
Boston, MA

Consultants and Advisors
Deborah Lowenberg Ball
Michael T. Battista
Marilyn Burns
Douglas H. Clements
Ann Grady

CONTENTS

TEACHER NOTES

WHERE TO START

The first-time user of *Number Games and Story Problems* should read the following:

When you next teach this same unit, you can begin to read more of the background. Each time you present the unit, you will learn more about how your students understand the mathematical ideas.

ABOUT THE *INVESTIGATIONS* CURRICULUM

Investigations in Number, Data, and Space® is a K–5 mathematics curriculum with four major goals:

- to offer students meaningful mathematical problems
- to emphasize depth in mathematical thinking rather than superficial exposure to a series of fragmented topics
- to communicate mathematics content and pedagogy to teachers
- to substantially expand the pool of mathematically literate students

The *Investigations* curriculum embodies an approach radically different from the traditional textbook-based curriculum. At each grade level, it consists of a set of separate units, each offering 2–8 weeks of work. These units of study are presented through investigations that involve students in the exploration of major mathematical ideas.

Approaching the mathematics content through investigations helps students develop flexibility and confidence in approaching problems, fluency in using mathematical skills and tools to solve problems, and proficiency in evaluating their solutions. Students also build a repertoire of ways to communicate about their mathematical thinking, while their enjoyment and appreciation of mathematics grow.

The investigations are carefully designed to invite all students into mathematics—girls and boys, members of diverse cultural, ethnic, and language groups, and students with different strengths and interests. Problem contexts often call on students to share experiences from their family, culture, or community. The curriculum eliminates barriers—such as work in isolation from peers, or emphasis on speed and memorization—that exclude some students from participating successfully in mathematics. The following aspects of the curriculum ensure that all students are included in significant mathematics learning:

- Students spend time exploring problems in depth.
- They find more than one solution to many of the problems they work on.
- They invent their own strategies and approaches, rather than relying on memorized procedures.
- They choose from a variety of concrete materials and appropriate technology, including calculators, as a natural part of their everyday mathematical work.
- They express their mathematical thinking through drawing, writing, and talking.
- They work in a variety of groupings—as a whole class, individually, in pairs, and in small groups.
- They move around the classroom as they explore the mathematics in their environment and talk with their peers.

While reading and other language activities are typically given a great deal of time and emphasis in elementary classrooms, mathematics often does not get the time it needs. If students are to experience mathematics in depth, they must have enough time to become engaged in real mathematical problems. We believe that a minimum of 5 hours of mathematics classroom time a week—about an hour a day—is critical at the elementary level. The plan and pacing of the *Investigations* curriculum are based on that belief.

We explain more about the pedagogy and principles that underlie these investigations in Teacher Notes throughout the units. For correlations of the curriculum to the NCTM Standards and further help in using this research-based program for teaching mathematics, see the following books:

- *Implementing the* Investigations in Number, Data, and Space® *Curriculum*
- *Beyond Arithmetic: Changing Mathematics in the Elementary Classroom* by Jan Mokros, Susan Jo Russell, and Karen Economopoulos

This book is one of the curriculum units for *Investigations in Number, Data, and Space.* In addition to providing part of a complete mathematics curriculum for your students, this unit offers information to support your own professional development. You, the teacher, are the person who will make this curriculum come alive in the classroom; the book for each unit is your main support system.

Although the curriculum does not include student textbooks, reproducible sheets for student work are provided in the unit and are also available as Student Activity Booklets. Students work actively with objects and experiences in their own environment and with a variety of manipulative materials and technology, rather than with a book of instruction and problems. We strongly recommend use of the overhead projector as a way to present problems, to focus group discussion, and to help students share ideas and strategies.

Ultimately, every teacher will use these investigations in ways that make sense for his or her particular style, the particular group of students, and the constraints and supports of a particular school environment. Each unit offers information and guidance for a wide variety of situations, drawn from our collaborations with many teachers and students over many years. Our goal in this book is to help you, a professional educator, implement this curriculum in a way that will give all your students access to mathematical power.

Investigation Format

The opening two pages of each investigation help you get ready for the work that follows.

What Happens This gives a synopsis of each session or block of sessions.

Mathematical Emphasis This lists the most important ideas and processes students will encounter in this investigation.

What to Plan Ahead of Time These lists alert you to materials to gather, sheets to duplicate, transparencies to make, and anything else you need to do before starting.

INVESTIGATION 3

Addition and Subtraction

What Happens

Session 1: Combining Situations Students extend their understanding of combining problems, in which they find the total of two amounts. They record and share their solution strategies.

Session 2: Separating Situations Students extend their understanding of separating problems, in which they find the result when one quantity is removed from another. They record and share their solution strategies.

Sessions 3, 4, and 5: Five-in-a-Row and Story Problems Students play a challenging version of Five-in-a-Row, a game introduced in an earlier unit that provides practice with single-digit addition pairs, this time with sums up to 20. They play this game and solve story problems for Choice Time. At the end of Choice Time, students share strategies for some of the story problems they solved.

Sessions 6, 7, and 8: Tens Go Fish In the game Tens Go Fish, students make combinations of ten with two addends. This game is added to the previous choices (Five-in-a-Row and story problems) as Choice Time continues for three more sessions. A whole-group round of Quick Images starts Session 7, and during the last half of Session 8, students share strategies for some of the story problems they solved.

Session 9: Combining with Unknown Change Students work with another type of story problem, "combining with unknown change" (a total and one of the amounts are given, and the second amount must be found). Again, they record and share their solution strategies. Then, for homework they write their own story problems to match a given addition expression.

Sessions 10, 11, and 12: Addition and Subtraction In the unit's final Choice Time, students work on Total of 20, (a variation of a game from Investigation 1), Tens Go Fish, and story problems. During the last half of Session 12, students share strategies for solving story problems.

Session 13: Solving Story Problems As an assessment, students solve a variety of combining and separating story problems and record their solution strategies.

Routines Refer to the section About Classroom Routines (pp. 166–173) for suggestions on integrating into the school day regular practice of mathematical skills in counting, exploring data, and understanding time and changes.

Mathematical Emphasis

- Visualizing combining and separating problem situations
- Developing strategies for solving combining and separating story problems
- Recording strategies for solving combining and separating story problems, using pictures, numbers, words, and equations
- Becoming familiar with combinations of 10 and 20
- Reasoning about combinations of 10
- Increasing familiarity with single-digit addition pairs

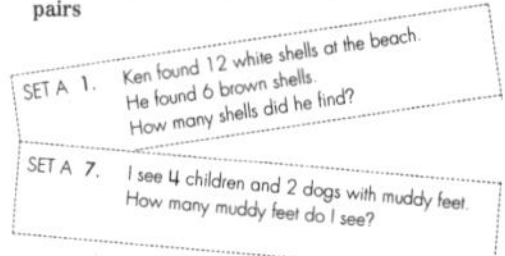

100 ▪ *Investigation 3: Addition and Subtraction*

INVESTIGATION 3

What to Plan Ahead of Time

Materials

- Number Cards: 1 deck per pair (Sessions 3–8, 10–12)
- Overhead projector (Sessions 3–8)
- Quick Image transparencies, Squares or Dot Addition Cards (Sessions 6–8)
- Interlocking cubes (Sessions 9–13)
- Paste or glue sticks (Sessions 3–8, 10–13)
- Counters, such as buttons, bread tabs, or pennies: at least 40 per pair (available)
- Unlined paper (available)
- Chart paper or newsprint (18 by 24 inches): 15–20 sheets (available)
- Envelopes for story problems (at least 21)

Other Preparation

- Duplicate student sheets and teaching resources, located at the end of this unit. If you have Student Activity Booklets, copy only items marked with an asterisk.

  For Session 1

  Student Sheet 19, At the Beach (p. 212): 1 per student, homework

  For Session 2

  Student Sheet 20, Clay Animals (p. 213): 1 per student, homework

  For Sessions 3, 4, and 5

  Story Problems, Set A (p. 220): 1 per student and 1 extra set. Cut apart and sort into seven envelopes. Paste one copy of the problem on the envelope for identification.

  Student Sheet 21, Five-in-a-Row with Three Cards (p. 214): 1 per student, homework

  Student Sheets 22–24, Five-in-a-Row Boards A–C (pp. 215–217): 1 of each per pair and a few extras (class), plus 1 of each per student, homework. Prepare a transparency of Board A.*

  For Sessions 6, 7, and 8

  Story Problems, Set B (p. 221): 1 per student and 1 extra set. Prepare like Set A.

  Student Sheet 25, Tens Go Fish (p. 218): 1 per student, homework

  For Session 9

  Student Sheet 26, Write Your Own Story Problem (p. 219): 1 per student, homework

  For Sessions 10, 11, and 12

  Story Problems, Set C (p. 222): 1 per student and 1 extra set. Prepare like Set A.

  Story Problems, Set D* (Challenges) (p. 223): enough for about half the class (optional)

  For Session 13

  Story Problems, Set E (p. 224): 1 per student. Cut apart and clip in five sets.
- On chart paper, write a combining story problem and a separating story problem for use in Sessions 1 and 2. Read through these sessions for information on choosing appropriate problems. On a third sheet, write a story problem that involves combining with unknown change (see p. 138) for Session 9.
- If your Number Cards are duplicated on paper, make "card holders" for the game Tens Go Fish. Fold a letter-size sheet in half the long way. With the fold at the bottom edge, fold up about 1 inch and staple ends to form a pocket. See illustration p. 132. (Sessions 6–8)

Investigation 3: Addition and Subtraction ▪ 101

Sessions Within an investigation, the activities are organized by class session, a session being at least a one-hour math class. Sessions are numbered consecutively through an investigation. Often several sessions are grouped together, presenting a block of activities with a single major focus.

> **Session 1**
>
> ## Combining Situations
>
> **Materials**
>
> - Combining problem on chart paper
> - Unlined paper
> - Student Sheet 19 (1 per student, homework)
>
> **What Happens**
>
> Students extend their understanding of combining problems, in which they find the total of two amounts. They record and share their solution strategies. Their work focuses on:
>
> - visualizing what happens in combining situations
> - understanding that when two amounts are combined, the result is more than the initial amounts
> - developing strategies for solving combining story problems
> - recording strategies for solving combining story problems, using pictures, numbers, words, and equations
>
> **Activity**
>
> **Making Sense of Combining**
>
> **Note:** If your students have worked in the unit *Building Number Sense,* they will have had experience with both combining and separating problems. First graders need many opportunities to develop their understanding of story problems and to learn ways of recording their thinking clearly.
>
> In this first activity, students solve story problems about combining by visualizing the amounts and the result. See the **Teacher Note,** Types of Story Problems (p. 108), for a discussion of the problem types students will be working with. Like the story problems in Investigation 2, the problems in this session involve finding a total. In Investigation 2, students found the total of several equal sets or groups: the number of hands in a group of people, or the number of wheels on several cars. In this session, the problems involve a sequence of actions in which two quantities are combined. Students need to figure out what is happening in the story: What does each amount represent? Are amounts being combined or separated? Will the result be more or less than the initial amount?
>
> Interpreting story problems that describe a sequence of actions can be challenging for first graders. In this whole-class activity, the numbers in the problems are deliberately kept small so that students can work mentally and can focus on the meaning of the story problem. In the next activity, Recording Combining Strategies, students will continue to work on making sense of combining story problems and they will record their solution strategies.
>
> **102** ▪ *Investigation 3: Addition and Subtraction*

When you find a block of sessions presented together—for example, Sessions 1, 2, and 3—read through the entire block first to understand the overall flow and sequence of the activities. Make some preliminary decisions about how you will divide the activities into three sessions for your class, based on what you know about your students. You may need to modify your initial plans as you progress through the activities, and you may want to make notes in the margins of the pages as reminders for the next time you use the unit.

Be sure to read the Session Follow-Up section at the end of the session block to see what homework assignments and extensions are suggested as you make your initial plans.

While you may be used to a curriculum that tells you exactly what each class session should cover, we have found that the teacher is in a better position to make these decisions. Each unit is flexible and may be handled somewhat differently by every teacher. While we provide guidance for how many sessions a particular group of activities is likely to need, we want you to be active in determining an appropriate pace and the best transition points for your class. It is not unusual for a teacher to spend more or less time than is proposed for the activities.

Activities The activities include pair and small-group work, individual tasks, and whole-class discussions. In any case, students are seated together, talking and sharing ideas during all work times. Students most often work cooperatively, although each student may record work individually.

Choice Time In most units, some sessions are structured with activity choices. In these cases, students may work simultaneously on different activities focused on the same mathematical ideas. Students choose which activities they want to do, and they cycle through them. You will need to decide how to set up and introduce these activities and how to let students make their choices. Some teachers set up choices as stations around the room, while others post the list of available choices and allow students to collect their own materials and choose their own work space. You may need to experiment with a few different structures before finding a setup that works best for you.

Extensions These follow-up activities are opportunities for some or all students to explore a topic in greater depth or in a different context. They are not designed for "fast" students; mathematics is a multifaceted discipline, and different students will want to go further in different investigations. Look for and encourage the sparks of interest and enthusiasm you see in your students, and use the extensions to help them pursue these interests.

Excursions Some of the *Investigations* units include excursions—blocks of activities that could be omitted without harming the integrity of the unit. This is one way of dealing with the great depth and variety of elementary mathematics—much more than a class has time to explore in any one year. Excursions give you the flexibility to make different choices from year to year, doing the excursion in one unit this time, and next year trying another excursion.

Tips for the Linguistically Diverse Classroom At strategic points in each unit, you will find concrete suggestions for simple modifications of the teaching strategies to encourage the participation of all students. Many of these tips offer alternative ways to elicit critical thinking from students at varying levels of English proficiency, as well as from other students who find it difficult to verbalize their thinking.

The tips are supported by suggestions for specific vocabulary work to help ensure that all students can participate fully in the investigations. The Preview for the Linguistically Diverse Classroom (p. I-21) lists important words that are assumed as part of the working vocabulary of the unit. Second-language learners will need to become familiar with these words in order to understand the problems and activities they will be doing. These terms can be incorporated into students' second-language work before or during the unit. Activities that can be used to present the words are found in the appendix, Vocabulary Support for Second-Language Learners (p. 174). In addition, ideas for making connections to students' language and cultures, included on the Preview page, help the class explore the unit's concepts from a multicultural perspective.

Classroom Routines Activities in counting, exploring data, and understanding time and changes are suggested for routines in the grade 1 *Investigations* curriculum. Routines offer ongoing work with this important content as a regular part of the school day. Some routines provide more practice with content presented in the curriculum; others extend the curriculum; still others explore new content areas.

Plan to incorporate a few of the routine activities into a standard part of your daily schedule, such as morning meeting. When opportunities arise, you can also include routines as part of your work in other subject areas (for example, keeping a weather chart for science). Most routines are short and can be done whenever you have a spare 10–15 minutes, such as before lunch or recess or at the end of the day.

You will need to decide how often to present routines, what variations are appropriate for your class, and at what points in the day or week you will include them. A reminder about classroom routines is included on the first page of each investigation. Whatever routines you choose, your students will gain the most from these routines if they work with them regularly.

Materials

A complete list of the materials needed for teaching this unit is found on p. I-17. Some of these materials are available in kits for the *Investigations* curriculum. Individual items can also be purchased from school supply dealers.

Classroom Materials In an active mathematics classroom, certain basic materials should be available at all times: interlocking cubes, pencils, unlined paper, graph paper, calculators, and things to count with. Some activities in this curriculum require scissors and glue sticks or tape. Stick-on notes and large paper are also useful materials

throughout. So that students can independently get what they need at any time, they should know where these materials are kept, how they are stored, and how they are to be returned to the storage area. Many teachers have found that stopping 5 minutes before the end of each session so that students can finish their work and clean up is helpful in maintaining classroom materials. You'll find that establishing such routines at the beginning of the year is well worth the time and effort.

Technology Calculators are introduced to students in the first unit of the grade 1 sequence, *Mathematical Thinking at Grade 1.* By freely exploring and experimenting, students become familiar with this important mathematical tool.

Computer activities at grade 1 use a software program, called *Shapes,* that was developed especially for the *Investigations* curriculum. This program is introduced in the geometry unit, *Quilt Squares and Block Towns.* Using *Shapes,* students explore two-dimensional geometry while making pictures and designs with pattern block shapes and tangram pieces.

Although the software is linked to activities only in the geometry unit, we recommend that students use it throughout the year. Thus, you may want to introduce it when you introduce pattern blocks in *Mathematical Thinking at Grade 1.* How you use the computer activities depends on the number of computers you have available. Suggestions are offered in the geometry unit for how to organize different types of computer environments.

Children's Literature Each unit offers a list of suggested children's literature (p. I-17) that can be used to support the mathematical ideas in the unit. Sometimes an activity is based on a specific children's book, with suggestions for substitutions where practical. While such activities can be adapted and taught without the book, the literature offers a rich introduction and should be used whenever possible.

Student Sheets and Teaching Resources Student recording sheets and other teaching tools needed for both class and homework are provided as reproducible blackline masters at the end of each

Name Date

Student Sheet 16

What's Missing? (A)

1		3		5		7		9	
11		13		15		17		19	
	22		24		26		28		30
	32		34		36		38		40
51	52		54	55		57	58		60
	62		64	65		67	68		70
	72			75			78		
81			84			87			90
91		93		95		97		99	

199

Investigation 2 • Sessions 10–12
Number Games and Story Problems

unit. They are also available as Student Activity Booklets. These booklets contain all the sheets each student will need for individual work, freeing you from extensive copying (although you may need or want to copy the occasional teaching resource on transparency film or card stock, or make extra copies of a student sheet).

We think it's important that students find their own ways of organizing and recording their work. They need to learn how to explain their thinking with both drawings and written words, and how to organize their results so someone else can understand them. For this reason, we deliberately do not provide student sheets for every activity. Regardless of the form in which students do their work, we recommend that they keep a mathematics notebook or folder so that their work is always available for reference.

Homework In *Investigations,* homework is an extension of classroom work. Sometimes it offers review and practice of work done in class, sometimes preparation for upcoming activities, and sometimes numerical practice that revisits work in earlier units. Homework plays a role both in supporting students' learning and in helping inform families about the ways in which students in this curriculum work with mathematical ideas.

Depending on your school's homework policies and your own judgment, you may want to assign more homework than is suggested in the units. For this purpose you might use the practice pages, included as blackline masters at the end of this unit, to give students additional work with numbers.

For some homework assignments, you will want to adapt the activity to meet the needs of a variety of students in your class: those with special needs, those ready for more challenge, and second-language learners. You might change the numbers in a problem, make the activity more or less complex, or go through a sample activity with those who need extra help. You can modify any student sheet for either homework or class use. In particular, making numbers in a problem smaller or larger can make the same basic activity appropriate for a wider range of students.

Another issue to consider is how to handle the homework that students bring back to class—how to recognize the work they have done at home without spending too much time on it. Some teachers hold a short group discussion of different approaches to the assignment; others ask students to share and discuss their work with a neighbor, or post the homework around the room and give students time to tour it briefly. If you want to keep track of homework students bring in, be sure it ends up in a designated place.

***Investigations* at Home** It is a good idea to make your policy on homework explicit to both students and their families when you begin teaching with *Investigations*. How frequently will you be assigning homework? When do you expect homework to be completed and brought back to school? What are your goals in assigning homework? How independent should families expect their children to be? What should the parent or guardian's role be? The more explicit you can be about your expectations, the better the homework experience will be for everyone.

Investigations at Home (a booklet available separately for each unit, to send home with students) gives you a way to communicate with families about the work students are doing in class. This booklet includes a brief description of every session, a list of the mathematics content emphasized in each investigation, and a discussion of each homework assignment to help families more effectively support their children. Whether or not you are using the *Investigations* at Home booklets, we expect you to make your own choices about homework assignments. Feel free to omit any and to add extra ones you think are appropriate.

Family Letter A letter that you can send home to students' families is included with the blackline masters for each unit. Families need to be informed about the mathematics work in your classroom; they should be encouraged to participate in and support their children's work. A reminder to send home the letter for each unit appears in one of the early investigations. These letters are also available separately in Spanish, Vietnamese, Cantonese, Hmong, and Cambodian.

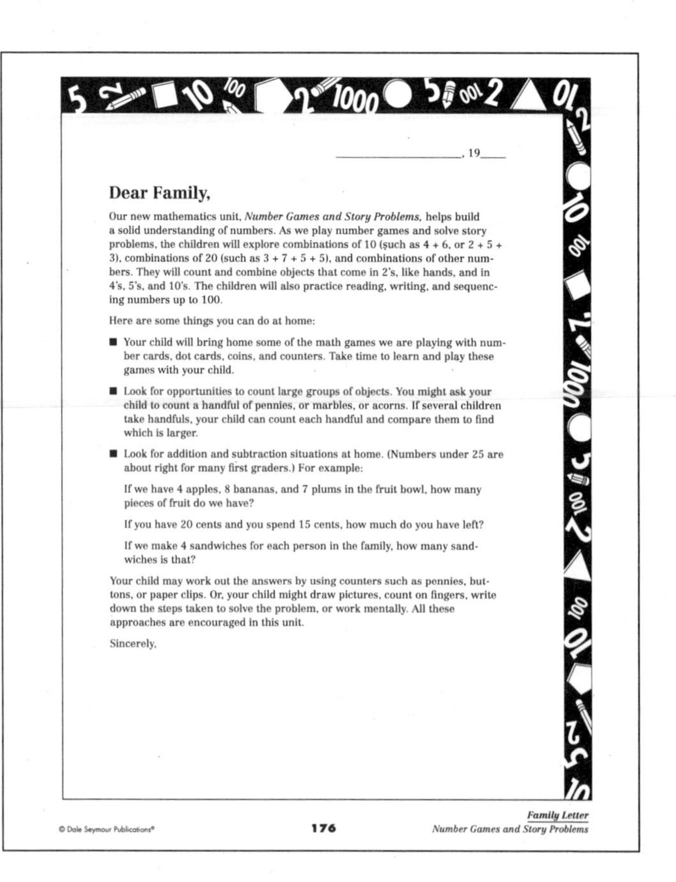

______________, 19____

Dear Family,

Our new mathematics unit, *Number Games and Story Problems,* helps build a solid understanding of numbers. As we play number games and solve story problems, the children will explore combinations of 10 (such as 4 + 6, or 2 + 5 + 3), combinations of 20 (such as 3 + 7 + 5 + 5), and combinations of other numbers. They will count and combine objects that come in 2's, like hands, and in 4's, 5's, and 10's. The children will also practice reading, writing, and sequencing numbers up to 100.

Here are some things you can do at home:

- Your child will bring home some of the math games we are playing with number cards, dot cards, coins, and counters. Take time to learn and play these games with your child.
- Look for opportunities to count large groups of objects. You might ask your child to count a handful of pennies, or marbles, or acorns. If several children take handfuls, your child can count each handful and compare them to find which is larger.
- Look for addition and subtraction situations at home. (Numbers under 25 are about right for many first graders.) For example:

  If we have 4 apples, 8 bananas, and 7 plums in the fruit bowl, how many pieces of fruit do we have?

  If you have 20 cents and you spend 15 cents, how much do you have left?

  If we make 4 sandwiches for each person in the family, how many sandwiches is that?

Your child may work out the answers by using counters such as pennies, buttons, or paper clips. Or, your child might draw pictures, count on fingers, write down the steps taken to solve the problem, or work mentally. All these approaches are encouraged in this unit.

Sincerely,

© Dale Seymour Publications® 176 *Family Letter* *Number Games and Story Problems*

Help for You, the Teacher

Because we believe strongly that a new curriculum must help teachers think in new ways about mathematics and about their students' mathematical thinking processes, we have included a great deal of material to help you learn more about both.

About the Mathematics in This Unit This introductory section (p. I-18) summarizes the critical information about the mathematics you will be teaching. It describes the unit's central mathematical ideas and how students will encounter them through the unit's activities.

Teacher Notes These reference notes provide practical information about the mathematics you are teaching and about our experience with how students learn. Many of the notes were written in response to actual questions from teachers, or to discuss important things we saw happening in the field-test classrooms. Some teachers like to read them all before starting the unit, then review them as they come up in particular investigations.

Dialogue Boxes Sample dialogues demonstrate how students typically express their mathematical ideas, what issues and confusions arise in their thinking, and how some teachers have guided class discussions. These dialogues are based on the extensive classroom testing of this curriculum; many are word-for-word transcriptions of recorded class discussions. They are not always easy reading; sometimes it may take some effort to unravel what the students are trying to say. But this is the value of these dialogues; they offer good clues to how your students may develop and express their approaches and strategies, helping you prepare for your own class discussions.

Where to Start You may not have time to read everything the first time you use this unit. As a first-time user, you will likely focus on understanding the activities and working them out with your students. Read completely through each investigation before starting to present it. Also read those sections listed in the Contents under the heading Where to Start (p. vi).

Teacher Note

Building on Number Combinations You Know

In first grade, most students are just becoming familiar with the addition combinations from 0 + 0 to 10 + 10. (See the **Teacher Note,** Strategies for Learning Addition Combinations, p. 159.) As they work with numbers in many different ways, some first graders begin to reason about the combinations they already know to figure out others. During this investigation, you may observe some of the following ways of thinking.

Counting On from a Known Combination When playing Dot Addition (p. 13), one student explained that she made 8 with 3 and 5 because "I know 3 and 3 is 6, and 2 more is 7, 8. But I didn't have enough 3's, so I put 2 and 3 together and made it 5." This approach is similar to *counting on,* a strategy that many students use for finding the total of two numbers. That is, to combine two numbers, they start at one and count up the other: "7 and 3 is — 8, 9, 10." Here, instead of beginning with a number, the student began with a number combination: 3 + 3. She counted up 2 from the total (6) to reach the goal number (8), and then increased one of the addends in the combination accordingly (3 + 5).

Counting Back from a Known Combination One student used a counting back approach when he was playing Counters in a Cup (p. 36) with 15 counters. His partner had hidden some of the 15 counters in a cup, and there were 6 remaining. This student reasoned that 9 were hidden because "10 plus 6 is 16, so take away 1 from 10 and it's 9, so 9 plus 6 is 15."

Probably your students will use counting back to find a new number combination only occasionally, and only with very familiar combinations. Counting backwards and building numbers by "taking away" is less familiar to first graders than counting up and building numbers by combining parts.

22 ▪ *Investigation 1: Number Combinations*

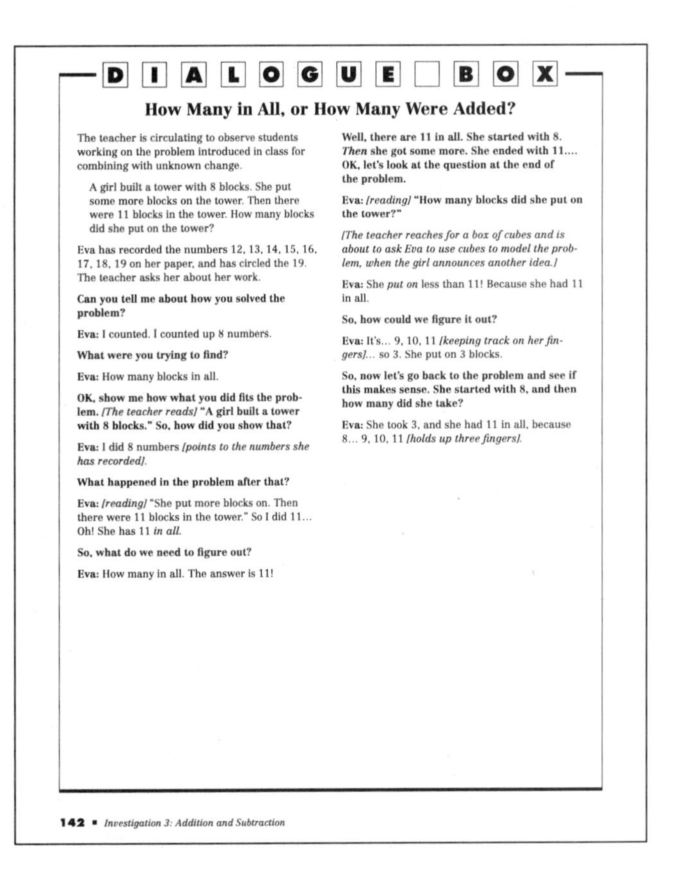

D I A L O G U E B O X

How Many in All, or How Many Were Added?

The teacher is circulating to observe students working on the problem introduced in class for combining with unknown change.

> A girl built a tower with 8 blocks. She put some more blocks on the tower. Then there were 11 blocks in the tower. How many blocks did she put on the tower?

Eva has recorded the numbers 12, 13, 14, 15, 16, 17, 18, 19 on her paper, and has circled the 19. The teacher asks her about her work.

Can you tell me about how you solved the problem?

Eva: I counted. I counted up 8 numbers.

What were you trying to find?

Eva: How many blocks in all.

OK, show me how what you did fits the problem. *[The teacher reads]* **"A girl built a tower with 8 blocks." So, how did you show that?**

Eva: I did 8 numbers *[points to the numbers she has recorded]*.

What happened in the problem after that?

Eva: *[reading]* "She put more blocks on. Then there were 11 blocks in the tower." So I did 11... Oh! She has 11 *in all.*

So, what do we need to figure out?

Eva: How many in all. The answer is 11!

Well, there are 11 in all. She started with 8. *Then* she got some more. She ended with 11.... OK, let's look at the question at the end of the problem.

Eva: *[reading]* **"How many blocks did she put on the tower?"**

[The teacher reaches for a box of cubes and is about to ask Eva to use cubes to model the problem, when the girl announces another idea.]

Eva: She *put on* less than 11! Because she had 11 in all.

So, how could we figure it out?

Eva: It's... 9, 10, 11 *[keeping track on her fingers]*... so 3. She put on 3 blocks.

So, now let's go back to the problem and see if this makes sense. She started with 8, and then how many did she take?

Eva: She took 3, and she had 11 in all, because 8... 9, 10, 11 *[holds up three fingers]*.

142 ▪ *Investigation 3: Addition and Subtraction*

The *Investigations* curriculum incorporates the use of two forms of technology in the classroom: calculators and computers. Calculators are assumed to be standard classroom materials, available for student use in any unit. Computers are explicitly linked to one or more units at each grade level; they are used with the unit on 2-D geometry unit at each grade, as well as with some of the units on measuring, data, and changes.

Using Calculators

In this curriculum, calculators are considered tools for doing mathematics, similar to pattern blocks or interlocking cubes. Just as with other tools, students must learn both *how* to use calculators correctly and *when* they are appropriate to use. This knowledge is crucial for daily life, as calculators are now a standard way of handling numerical operations, both at work and at home.

Using a calculator correctly is not a simple task; it depends on a good knowledge of the four operations and of the number system, so that students can select suitable calculations and also determine what a reasonable result would be. These skills are the basis of any work with numbers, whether or not a calculator is involved.

Unfortunately, calculators are often seen as tools to check computations with, as if other methods are somehow more fallible. Students need to understand that any computational method can be used to check any other; it's just as easy to make a mistake on the calculator as it is to make a mistake on paper or with mental arithmetic. Throughout this curriculum, we encourage students to solve computation problems in more than one way in order to double-check their accuracy. We present mental arithmetic, paper-and-pencil computation, and calculators as three possible approaches.

In this curriculum we also recognize that, despite their importance, calculators are not always appropriate in mathematics instruction. Like any tools, calculators are useful for some tasks, but not for others. You will need to make decisions about when to allow students access to calculators and when to ask that they solve problems without them, so that they can concentrate on other tools and skills. At times when calculators are or are not appropriate for a particular activity, we make specific recommendations. Help your students develop their own sense of which problems they can tackle with their own reasoning and which ones might be better solved with a combination of their own reasoning and the calculator.

Managing calculators in your classroom so that they are a tool, and not a distraction, requires some planning. When calculators are first introduced, students often want to use them for everything, even problems that can be solved quite simply by other methods. However, once the novelty wears off, students are just as interested in developing their own strategies, especially when these strategies are emphasized and valued in the classroom. Over time, students will come to recognize the ease and value of solving problems mentally, with paper and pencil, or with manipulatives, while also understanding the power of the calculator to facilitate work with larger numbers.

Experience shows that if calculators are available only occasionally, students become excited and distracted when they are permitted to use them. They focus on the tool rather than on the mathematics. In order to learn when calculators are appropriate and when they are not, students must have easy access to them and use them routinely in their work.

If you have a calculator for each student, and if you think your students can accept the responsibility, you might allow them to keep their calculators with the rest of their individual materials, at least for the first few weeks of school. Alternatively, you might store them in boxes on a shelf, number each calculator, and assign a corresponding number to each student. This system can give students a sense of ownership while also helping you keep track of the calculators.

Using Computers

Students can use computers to approach and visualize mathematical situations in new ways. The computer allows students to construct and manipulate geometric shapes, see objects move according to rules they specify, and turn, flip, and repeat a pattern.

This curriculum calls for computers in units where they are a particularly effective tool for learning mathematics content. One unit on 2-D geometry at each of the grades 3–5 includes a core of activities that rely on access to computers, either in the classroom or in a lab. Other units on geometry, measurement, data, and changes include computer activities, but can be taught without them. In these units, however, students' experience is greatly enhanced by computer use.

The following list outlines the recommended use of computers in this curriculum:

Grade 1
Unit: *Survey Questions and Secret Rules* (Collecting and Sorting Data)
Software: Tabletop, Jr.
Source: Broderbund

Unit: *Quilt Squares and Block Towns* (2-D and 3-D Geometry)
Software: *Shapes*
Source: provided with the unit

Grade 2
Unit: *Mathematical Thinking at Grade 2* (Introduction)
Software: *Shapes*
Source: provided with the unit

Unit: *Shapes, Halves, and Symmetry* (Geometry and Fractions)
Software: *Shapes*
Source: provided with the unit

Unit: *How Long? How Far?* (Measuring)
Software: *Geo-Logo*
Source: provided with the unit

Grade 3
Unit: *Flips, Turns, and Area* (2-D Geometry)
Software: *Tumbling Tetrominoes*
Source: provided with the unit

Unit: *Turtle Paths* (2-D Geometry)
Software: *Geo-Logo*
Source: provided with the unit

Grade 4
Unit: *Sunken Ships and Grid Patterns* (2-D Geometry)
Software: *Geo-Logo*
Source: provided with the unit

Grade 5
Unit: *Picturing Polygons* (2-D Geometry)
Software: *Geo-Logo*
Source: provided with the unit

Unit: *Patterns of Change* (Tables and Graphs)
Software: *Trips*
Source: provided with the unit

Unit: *Data: Kids, Cats, and Ads* (Statistics)
Software: Tabletop, Sr.
Source: Broderbund

The software provided with the *Investigations* units uses the power of the computer to help students explore mathematical ideas and relationships that cannot be explored in the same way with physical materials. With the *Shapes* (grades 1–2) and *Tumbling Tetrominoes* (grade 3) software, students explore symmetry, pattern, rotation and reflection, area, and characteristics of 2-D shapes. With the *Geo-Logo* software (grades 3–5), students investigate rotations and reflections, coordinate geometry, the properties of 2-D shapes, and angles. The *Trips* software (grade 5) is a mathematical exploration of motion in which students run experiments and interpret data presented in graphs and tables.

We suggest that students work in pairs on the computer; this not only maximizes computer resources but also encourages students to consult, monitor, and teach one another. Generally, more than two students at one computer find it difficult to share. Managing access to computers is an issue for every classroom. The curriculum gives you explicit support for setting up a system. The units are structured on the assumption that you have enough computers for half your students to work on the machines in pairs at one time. If you do not have access to that many computers, suggestions are made for structuring class time to use the unit with five to eight computers, or even with fewer than five.

Assessment plays a critical role in teaching and learning, and it is an integral part of the *Investigations* curriculum. For a teacher using these units, assessment is an ongoing process. You observe students' discussions and explanations of their strategies on a daily basis and examine their work as it evolves. While students are busy recording and representing their work, working on projects, sharing with partners, and playing mathematical games, you have many opportunities to observe their mathematical thinking. What you learn through observation guides your decisions about how to proceed. In any of the units, you will repeatedly consider questions like these:

- Do students come up with their own strategies for solving problems, or do they expect others to tell them what to do? What do their strategies reveal about their mathematical understanding?
- Do students understand that there are different strategies for solving problems? Do they articulate their strategies and try to understand other students' strategies?
- How effectively do students use materials as tools to help with their mathematical work?
- Do students have effective ideas for keeping track of and recording their work? Does keeping track of and recording their work seem difficult for them?

You will need to develop a comfortable and efficient system for recording and keeping track of your observations. Some teachers keep a clipboard handy and jot notes on a class list or on adhesive labels that are later transferred to student files. Others keep loose-leaf notebooks with a page for each student and make weekly notes about what they have observed in class.

Assessment Tools in the Unit

With the activities in each unit, you will find questions to guide your thinking while observing the students at work. You will also find two built-in assessment tools: Teacher Checkpoints and embedded Assessment activities.

Teacher Checkpoints The designated Teacher Checkpoints in each unit offer a time to "check in" with individual students, watch them at work, and ask questions that illuminate how they are thinking.

At first it may be hard to know what to look for, hard to know what kinds of questions to ask. Students may be reluctant to talk; they may not be accustomed to having the teacher ask them about their work, or they may not know how to explain their thinking. Two important ingredients of this process are asking students open-ended questions about their work and showing genuine interest in how they are approaching the task. When students see that you are interested in their thinking and are counting on them to come up with their own ways of solving problems, they may surprise you with the depth of their understanding.

Teacher Checkpoints also give you the chance to pause in the teaching sequence and reflect on how your class is doing overall. Think about whether you need to adjust your pacing: Are most students fluent with strategies for solving a particular kind of problem? Are they just starting to formulate good strategies? Or are they still struggling with how to start? Depending on what you see as the students work, you may want to spend more time on similar problems, change some of the problems to use smaller numbers, move quickly to more challenging material, modify subsequent activities for some students, work on particular ideas with a small group, or pair students who have good strategies with those who are having more difficulty.

Embedded Assessment Activities Assessment activities embedded in each unit will help you examine specific pieces of student work, figure out what it means, and provide feedback. From the students' point of view, these assessment activities are no different from any others. Each is a learning experience in and of itself, as well as an opportunity for you to gather evidence about students' mathematical understanding.

The embedded assessment activities sometimes involve writing and reflecting; at other times, a discussion or brief interaction between student and teacher; and in still other instances, the creation and explanation of a product. In most cases, the assessments require that students *show* what they did, *write* or *talk* about it, or do both. Having to explain how they worked through a problem helps students be more focused and clear in their mathematical thinking. It also helps them realize that doing mathematics is a process that may involve tentative starts, revising one's approach, taking different paths, and working through ideas.

Teachers often find the hardest part of assessment to be interpreting their students' work. We provide guidelines to help with that interpretation. If you have used a process approach to teaching writing, the assessment in *Investigations* will seem familiar. For many of the assessment activities, a Teacher Note provides examples of student work and a commentary on what it indicates about student thinking.

Documentation of Student Growth

To form an overall picture of mathematical progress, it is important to document each student's work in journals, notebooks, or portfolios. The choice is largely a matter of personal preference; some teachers have students keep a notebook or folder for each unit, while others prefer one mathematics notebook, or a portfolio of selected work for the entire year. The final activity in each *Investigations* unit, called Choosing Student Work to Save, helps you and the students select representative samples for a record of their work.

This kind of regular documentation helps you synthesize information about each student as a mathematical learner. From different pieces of evidence, you can put together the big picture. This synthesis will be invaluable in thinking about where to go next with a particular child, deciding where more work is needed, or explaining to parents (or other teachers) how a child is doing.

If you use portfolios, you need to collect a good balance of work, yet avoid being swamped with an overwhelming amount of paper. Following are some tips for effective portfolios:

- Collect a representative sample of work, including some pieces that students themselves select for inclusion in the portfolio. There should be just a few pieces for each unit, showing different kinds of work—some assignments that involve writing, as well as some that do not.
- If students do not date their work, do so yourself so that you can reconstruct the order in which pieces were done.
- Include your reflections on the work. When you are looking back over the whole year, such comments are reminders of what seemed especially interesting about a particular piece; they can also be helpful to other teachers and to parents. Older students should be encouraged to write their own reflections about their work.

Assessment Overview

There are two places to turn for a preview of the assessment opportunities in each *Investigations* unit. The Assessment Resources column in the unit Overview Chart (pp. I-13–I-16) identifies the Teacher Checkpoints and Assessment activities embedded in each investigation, guidelines for observing the students that appear within classroom activities, and any Teacher Notes and Dialogue Boxes that explain what to look for and what types of student responses you might expect to see in your classroom. Additionally, the section About the Assessment in This Unit (p. I-20) gives you a detailed list of questions for each investigation, keyed to the mathematical emphases, to help you observe student growth.

Depending on your situation, you may want to provide additional assessment opportunities. Most of the investigations lend themselves to more frequent assessment, simply by having students do more writing and recording while they are working.

Number Games and Story Problems

Content of This Unit Students deepen their understanding of number in several ways: by finding combinations of 10 (such as 4 + 6, or 2 + 5 + 3), combinations of 20 (such as 3 + 7 + 5 + 5), and combinations of other numbers; by counting and combining different kinds of collections (sets of dot cards, things that come in 2's, 4's, 5's, and 10's, and sets of coins); and by reading, writing, and sequencing numbers up to 100. They use their growing understanding of number to solve a variety of addition and subtraction story problems, while they learn to recognize and interpret addition and subtraction situations, choose and carry out strategies for solving the problems, and record their solution strategies clearly. Sometimes students will solve problems by modeling them with objects or pictures; other times they will count; and other times they will begin using what they know about numbers and number relationships. All these approaches are encouraged in this unit.

Connections with Other Units If you are doing the full-year *Investigations* curriculum, this is the fifth of six units. It offers a direct extension of concepts presented in the unit *Building Number Sense,* including number combinations, counting, and developing and recording strategies for solving story problems. If your students have worked with *Building Number Sense,* they will already be familiar with some of the games and activities in this unit, but now they will be working with larger numbers and more challenging versions.

This unit can also be used successfully at grade 2, depending on the previous experience and needs of your students.

Investigations Curriculum ▪ Suggested Grade 1 Sequence

Mathematical Thinking at Grade 1 (Introduction)

Building Number Sense (The Number System)

Survey Questions and Secret Rules (Collecting and Sorting Data)

Quilt Squares and Block Towns (2-D and 3-D Geometry)

▶ *Number Games and Story Problems* (Addition and Subtraction)

Bigger, Taller, Heavier, Smaller (Measuring)

Investigation 1 ▪ Number Combinations

Class Sessions	Activities	Pacing
Session 1 (p. 5) PICTURES OF 10	Quick Image Pictures of 10 Homework	minimum 1 hr
Sessions 2 and 3 (p. 9) NUMBER COMBINATIONS	Dot Addition On and Off Choice Time Quick Image Dot Combinations Homework: Dot Addition, On and Off Extensions: Most and Fewest, Combinations of One Number	minimum 2 hr
Sessions 4 and 5 (p. 17) TOTAL OF 10	Introducing Total of 10 Choice Time Sharing Dot Addition Sums Homework: Total of 10	minimum 2 hr
Session 6 (p. 24) HOW MANY OF EACH COLOR?	Teacher Checkpoint: Ten Crayons Sharing Solutions Quick Image Dot Combinations Extensions: Combining Solutions to Make 20, Fifteen Crayons, Crayons in Four Colors, Ways to Make 10	minimum 1 hr
Sessions 7, 8, and 9 (p. 30) CRAYON PUZZLES	Introducing Crayon Puzzles Counters in a Cup Choice Time Sharing Combinations of 10 Homework: Counters in a Cup, Math Games Extension: Finding All the Combinations of 10	minimum 3 hr
Session 10 (Excursion)* (p. 40) NUMBER COMBINATION STORIES	Reading a Number Combination Book Creating Number Combination Stories Extension: All the Combinations	minimum 1 hr
Classroom Routines (see pp. 166–173)		

*Excursions can be omitted without harming the integrity or continuity of the unit but offer good mathematical work if you have time to include them.

Mathematical Emphasis

- Finding combinations of numbers up to about 20
- Finding the total of two or more single-digit numbers
- Exploring relationships among different combinations of a number

Assessment Resources

Observing the Students (pp. 14, 20, 26, 36, 44)

Pictures and Equations (Dialogue Box, p. 8)

Building on Number Combinations You Know (Teacher Note, p. 22)

Teacher Checkpoint: Ten Crayons (p. 24)

Combinations of 10 (Dialogue Box, p. 39)

Materials

Overhead projector
Number Cards
Cubes, counters, crayons
Empty crayon box
Paper cups
Ten Flashing Fireflies by Philemon Sturges (opt.)
Chart paper or newsprint
Unlined paper
Envelopes
Resealable plastic bags
Student Sheets 1–9
Teaching resource sheets

Investigation 2 ▪ Twos, Fives, and Tens

Class Sessions	Activities	Pacing
Session 1 (p. 49) HOW MANY HANDS?	Hands and Other Pairs How Many Hands? Sharing Strategies Counting Hands in the Group Homework: How Many Hands at Home? Extensions: How Many Fingers in the Group?, How Many Feet in the Bed?, Counting Around the Class by 2's	minimum 1 hr
Session 2 (p. 59) TWOS AND FOURS	How Many Hands at Home? Patterns in People and Hands A Problem About Fours Homework: Cats and Paws Extensions: What Comes in Twos?, What Else Has As Many Hands?	minimum 2 hr
Session 3 (p. 64) COLLECT 25¢ TOGETHER	Exploring Coins Collect 25¢ Together Homework: Collect 25¢ Together Extension: Ways to Make 15¢	minimum 2 hr
Sessions 4 and 5 (p. 70) COUNTING AND COMBINING	Quick Images with Squares Introducing How Many Squares? Choice Time Sharing Solution Strategies Extension: Squares in Threes and Sixes	minimum 2 hr
Sessions 6, 7, and 8 (p. 78) NUMBERS TO 100	Exploring the 100 Chart Missing Numbers Choice Time Quick Image Squares How We Counted Homework: Coins	minimum 3 hr
Session 9 (p. 84) PATTERNS OF FIVES AND TENS	Playing Missing Numbers Together Clapping Patterns A Pattern with Five Parts	minimum 1 hr
Sessions 10, 11, and 12 (p. 88) TWOS, FIVES, AND TENS	Introducing Roll Tens Choice Time (with Assessment: How Many Squares?) Homework: What's Missing? Extensions: How Many More Cubes?, Counting on the Calculator, Collections of 100, 100 Children	minimum 3 hr
Session 13 (Excursion)* (p. 98) COUNTING BY KANGAROOS	*Counting by Kangaroos* Finding the Total in Three Groups Extension: How Many Animals in All?	minimum 1 hr
Classroom Routines (see pp. 166–173)		

*Excursions can be omitted without harming the integrity or continuity of the unit, but offer good mathematical work if you have time to include them.

Continued on next page

Mathematical Emphasis

- Developing strategies for organizing sets of objects so that they are easy to count and combine
- Finding the total of several 2's, 4's, 5's or 10's
- Recording strategies for counting and combining, using pictures, numbers, and words
- Reading, writing, and sequencing numbers to 100
- Becoming familiar with coins and equivalencies among them

Assessment Resources

Observing the Students (pp. 50, 61, 66, 74, 82, 87, 92)

Helping Students Record Their Strategies (Dialogue Box, p. 57)

What Do You Notice About Coins? (Dialogue Box, p. 69)

Coin Equivalencies in Grade 1 (Teacher Note, p. 68)

Finding Ways to Count (Dialogue Box, p. 76)

Assessment: How Many Squares? (p. 91 and Teacher Note, p. 95)

Materials

Dot cubes
Coins (play or real)
Hundred Number Wall Chart
Plastic 100 Number Boards
Interlocking cubes
Overhead projector
Pattern blocks
Calculators
Counters
Envelopes for storage
Unlined paper
Chart paper or newsprint
Student Sheets 10–18
Teaching resource sheets

Investigation 3 ▪ Addition and Subtraction

Class Sessions	Activities	Pacing
Session 1 (p. 102) COMBINING SITUATIONS	Making Sense of Combining Recording Combining Strategies Sharing Combining Strategies Homework: At the Beach	minimum 1 hr
Session 2 (p. 109) SEPARATING SITUATIONS	Making Sense of Separating Recording Separating Strategies Sharing Separating Strategies Homework: Clay Animals	minimum 1 hr
Sessions 3, 4, and 5 (p. 117) FIVE-IN-A-ROW AND STORY PROBLEMS	Five-in-a-Row with Three Cards Choice Time Sharing Story Problem Strategies Homework: Story Problems, Five-in-a-Row with Three Cards	minimum 3 hr
Sessions 6, 7, and 8 (p. 129) TENS GO FISH	Introducing Tens Go Fish Teacher Checkpoint: Choice Time Quick Images Sharing Story Problem Strategies Homework: Story Problems, Tens Go Fish	minimum 3 hr
Session 9 (p. 138) COMBINING WITH UNKNOWN CHANGE	Combining with Unknown Change Writing a Story Problem Homework: Writing a Story Problem Extension: Combining and Separating Stories	minimum 1 hr
Sessions 10, 11, and 12 (p. 143) ADDITION AND SUBTRACTION	Choice Time Sharing Story Problem Strategies Homework: Story Problems, Math Games Extension: Close to 20	minimum 3 hr
Session 13 (p. 149) SOLVING STORY PROBLEMS	Assessment: Solving Story Problems Choosing Student Work to Save	minimum 1 hr
Classroom Routines (see pp. 166–173)		

Mathematical Emphasis

- Visualizing combining and separating problem situations
- Developing strategies for solving combining and separating story problems
- Recording strategies for solving combining and separating story problems, using pictures, numbers, words, and equations

Assessment Resources

Observing the Students (pp. 104, 121, 133, 146, 150)

Three Approaches to Story Problems (Teacher Note, p. 113)

How Many in All, or How Many Were Added? (Dialogue Box, p. 142)

Assessment: Solving Story Problems (p. 149 and Teacher Note, p. 154)

Materials

Number Cards
Overhead projector
Interlocking cubes
Counters
Chart paper or newsprint
Unlined paper
Paste or glue sticks
Student Sheets 19–26
Teaching resource sheets

Following are the basic materials needed for the activities in this unit. Many items can be purchased from the publisher, either individually or in the Teacher Resource Package and the Student Materials Kit for grade 1. Detailed information is available on the *Investigations* order form. To obtain this form, call toll-free 1-800-872-1100 and ask for a Dale Seymour customer service representative.

Dot cubes (2 per pair)

Hundred Number Wall Chart (vinyl, with pockets), with numeral cards and chart markers (translucent colored plastic overlays)

Plastic Hundred Number Boards with removable tiles (2–3 for the class)

Primary Number Cards: one deck per pair (manufactured or make your own)

Sets of coins, play or real: one set per pair with 30 pennies, 6–7 nickels, 3–4 dimes, and 1–2 quarters

Counters (buttons, bread tabs, pennies): at least 30 per student

Class sets of pattern blocks, interlocking cubes

Paper cups or other small containers (1 per student)

Calculators (at least 6–8 for the class)

Ten Flashing Fireflies by Philomen Sturges (optional)

Counting by Kangaroos by Joy N. Hulme (optional)

Crayons or markers

Unlined paper

Chart paper or newsprint (18 by 24 inches)

Envelopes for storing overhead transparencies

Resealable plastic bags for storing cards

Paste or glue sticks

Overhead projector and blank transparencies

Empty crayon box or other box that holds ten crayons, with an opening large enough that everyone can see the colors of the crayons inside

The following materials are provided at the end of this unit as blackline masters. A Student Activity Booklet containing all student sheets and teaching resources needed for individual work is available.

Family Letter (p. 176)

Student Sheets 1–26 (p. 177)

Teaching Resources:

Quick Image Pictures of 10 (p. 186)
Dot Addition Cards (p. 187)
Blank Dot Addition Board (p. 188)
Crayon Puzzles (p. 189)
Quick Image Squares (p. 202)
Squares (p. 205)
Blank 100 Chart (p. 208)
200 Chart (p. 209)
Roll Tens Game Mats (p. 210)
Story Problems (p. 220)
Number Cards (p. 225)

Practice Pages (p. 229)

Related Children's Literature

Aker, Suzanne. *What Comes in 2's, 3's, and 4's?* New York: Simon and Schuster Books for Young Readers, 1990.

Anno, Mitsumasa. *Anno's Counting House.* New York: Philomel Books, 1982.

Bogart, Jo Ellen. *Ten for Dinner.* New York: Scholastic, 1989.

Gordon, Jeffie Ross. *Six Sleepy Sheep.* Honesdale, Pennsylvania: Boyds Mills Press, Caroline House, 1991.

Hamm, Diane Johnson. *How Many Feet in the Bed?* New York: Simon and Schuster, 1991.

Hulme, Joy N. *Counting by Kangaroos.* New York: W.H. Freeman, 1995.

Jonas, Ann. *Splash!* New York: Greenwillow Books, 1985.

Kuskin, Karla. *The Philharmonic Gets Dressed.* New York: HarperCollins, 1992.

Sturges, Philemon. *Ten Flashing Fireflies.* New York: North-South Books, 1995.

In this unit, students learn about the number system and number relationships, and they use their growing understanding of number to solve a variety of addition and subtraction story problems. Students build on the concepts introduced in the unit *Building Number Sense* as they deepen and extend their understanding of number.

The Number System Counting is the basis for understanding our number system and for almost all of the number work primary grade students do. In first grade, students work on reading, writing, and sequencing numbers, counting orally, and counting quantities. They also learn about important relationships among the counting words, the written numbers, and the quantities they represent. For example, they begin to see that whenever they count up 1, they are referring to a quantity that has one more than the quantity they started with, and whenever they count down 1, they are referring to a quantity that has one less. By the end of the year, many students will have learned the oral counting sequence to 100, will recognize some patterns in the sequence of written numbers from 1 to 100, and will have a solid grasp of quantities up to about 25.

As students grow proficient at counting by 1's, they are ready to begin counting by numbers other than 1. This is a complicated idea that students come to grasp gradually over the early elementary years. In this unit, students are just beginning to make sense of this as they visualize quantities grouped in 2's, 5's, and 10's. For example, in the activity How Many Squares? they count collections of single squares and squares joined together in groups, as in this set:

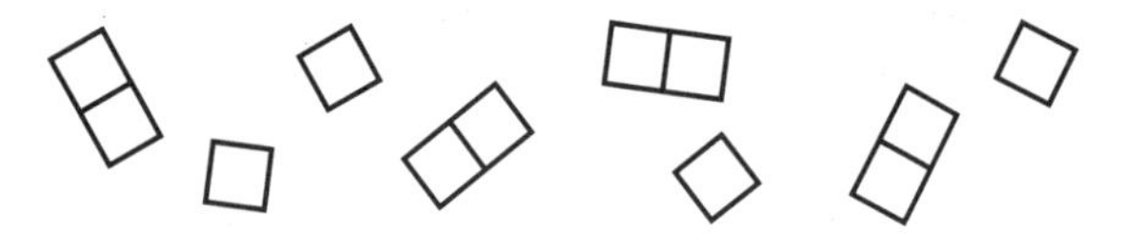

At first, students will likely find the total by counting each square. Many soon discover that they can organize the squares in some way to help them keep track as they count. For example, they might put the squares in groups of 2, 4, or 5, or they might put all the pairs in one row and all the single squares in another. This kind of organizing is an important step in beginning to make sense of units greater than 1. Even though they begin to organize in groups, many first graders will still need to count the squares in each group by 1. Others may begin to sum the sizes of the groups ("These two 2's are 4, 4 and 2 more is 6, 6 and 2 is 8..." and so on). Still others may count two or three pairs by 2's and then return to counting by 1's as quantities increase (2, 4, 6, 7, 8, 9, 10, 11, 12). Or, they may count the entire set by 2's, and then, if they do not completely trust the totals they get that way, they will count again by 1's to check. All are important ways in which students explore thinking about units greater than 1 and how these units are related to the 1's out of which they are constructed.

How Many Squares? gives students one model of building a quantity from equal-sized pieces. In other activities, students work with patterns of 5's and 10's on the 100 chart, interlocking cubes in rows of 10's, coin collections, and problems in which they find the total of several 2's, 4's, or 5's (for example, how many legs do four dogs have?). Again, some students will approach these activities with strategies that involve 1's, while others will begin developing strategies that involve counting by and combining numbers other than 1.

Number Relationships As students become more comfortable using one number to stand for a group of objects, they can start breaking numbers apart in different ways. In the early elementary grades, students develop a basic repertoire of number combinations, and they begin to use combinations they know to figure out others. One way that students build their knowledge of number combinations in this unit is by playing games in which they make totals in different ways. For example, in the game Dot Addition, students make given totals by combining cards with
2, 3, 4, or 5 dots.

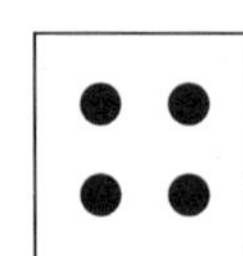
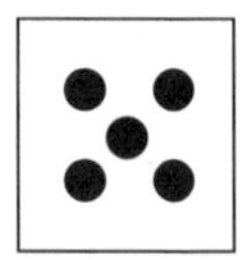

Initially, students might approach the task of making a total of 15 by combining cards and counting the dots until they find a set that works: "Here's a 5 and a 3 card. That's 1, 2, 3, 4, 5, 6, 7, 8. What if I use another 3 card? That's 9, 10, 11. I still need more." Over time, many students will begin to use number combinations they know to help them find their solutions: "I'll start with two 5 cards. Five and 5 is 10, and I need another—11, 12, 13, 14, 15—five more." By the end of the year, some first graders may find their solutions by reasoning about relationships among number combinations: "Fifteen is 8 and 7, so I'll break up 8 into 5 and 3, and 7 into 5 and 2. Oh, I've used up all my 2 cards, so I'll take 1 from the 5 and put it with the 2, and I'll make 7 with 4 and 3. So, that's 5 + 3 + 4 + 3." Through the next couple of years, students will continue to develop this kind of flexible thinking about number, and they will work on constructing addition and subtraction strategies based on this way of thinking.

Most of students' work with number combinations focuses on numbers 20 and under, with particular emphasis on combinations of 10. By the end of the year, many first graders will begin using some knowledge of number combinations to solve problems, especially those involving familiar numbers under 10 or 12. However, they will probably continue counting by 1's in many situations, particularly when larger numbers are involved.

Addition and Subtraction Consider the following story problem:

> I had 18 marbles. My sister gave me 8 marbles. How many marbles do I have now?

In order to solve this problem, students first need to make sense of the situation. What sequence of actions is being described in the problem? What does each amount represent? Is the second amount to be combined with the first, or is it to separated (or "taken away") from the first? Will the result be more or less than the initial amount?

Next, students need to find a way to solve the problem. Many first graders will use direct modeling or counting strategies: They might draw a picture of 18 marbles and a picture of 8 marbles, and then count them all. Or, they might count up 8 from 18. As their knowledge of number combinations and their understanding of the base ten structure of the number system grows throughout first and second grades, they will gradually develop strategies involving numerical reasoning. One student might explain that "I broke 8 into 2 and 6, counted up 2 from 18—19, 20, and then I just knew that 20 and 6 is 26." Another might say that "18 is 10 and 8, 8 and 8 is 16, and 10 more is 26."

All these approaches are appropriate for young students working with addition and subtraction problems. Some students will solve all addition and subtraction problems with strategies that involve counting by 1's until well into second grade. Some will begin using numerical reasoning, especially when combining small, very familiar numbers, but will return to direct modeling or counting strategies to combine larger numbers. A few students may consistently use strategies that involve number combinations and breaking numbers into convenient chunks, including multiples of ten and ones. It is essential that all students have the opportunity to construct approaches that are firmly grounded in their own developing number sense, that they can rely on to solve problems, and that they can explain clearly to others.

Another emphasis is on recording solution strategies clearly, using pictures, numbers, and words, as well as standard notation for addition and subtraction equations. Recording work is an important way to think through a problem carefully, to keep track of procedures, and to communicate solution methods to others. As students share work with you and with their classmates, they learn how to explain their thinking so that someone else can understand it. In the process, they become aware of different solution methods.

Mathematical Emphasis At the beginning of each investigation, the Mathematical Emphasis section tells you what is most important for students to learn about during that investigation. Many of these understandings and processes are difficult and complex. Students gradually learn more and more about each idea over many years of schooling. Individual students will begin and end the unit with different levels of knowledge and skill, but all will learn more about the number system and number relationships and will develop their understanding of addition and subtraction situations.

Throughout the *Investigations* curriculum, there are many opportunities for ongoing daily assessment as you observe, listen to, and interact with students at work. In this unit, you will find two Teacher Checkpoints:

Investigation 1, Session 6:
Ten Crayons (p. 24)

Investigation 3, Session 6–8:
Choice Time—Story Problems (p. 131)

This unit also has two embedded assessment activities:

Investigation 2, Sessions 10–12:
Choice Time—How Many Squares? (p. 91)

Investigation 3, Session 13:
Solving Story Problems (p. 149)

In addition, you can use almost any activity in this unit to assess your students' needs and strengths. Listed below are questions to help you focus your observations in each investigation. You may want to keep track of your observations for each student to help you plan your curriculum and monitor students' growth.

Investigation 1: Number Combinations

- How do students generate combinations of numbers equal to a given total? Do they take a random approach, combining numbers until they reach the total? Do they use strategies based on counting? Do they use strategies based on number combinations?

- How do students find the total of two or more numbers? Do they count from 1? Do they count on from a number? Do they use knowledge of number combinations? Do they reason from number combinations they know to find other combinations? Can they quickly add on 1 or 2 to a number, or do they need to count out each quantity?

- How do students find more than one combination of a number? Do they treat each combination as a separate problem? Do they find new combinations by changing something about a combination they've already found? Do they keep track of combinations they've already found to make sure they don't duplicate?

Investigation 2: Twos, Fives, and Tens

- When counting a set of objects, do students organize them in some way, or do they count in a random order? Do they arrange them in equal groups, such as groups of two or five?

- What strategies do students use for finding a total of several equal amounts? Do they count by 1's? Do they count by another number? Do they use strategies that involve number combinations? doubles? tens and ones?

- Can students record their counting and combining strategies clearly? Do they use pictures? words? How are they using numbers or equations as part of their recording?

- What range of numbers are students comfortable reading, writing, and sequencing? Do students use patterns in the number sequence to help them?

- What coins are students familiar with? Do they know coin names? values? Can students count a set of different kinds of coins accurately? Can they make trades among coins?

Investigation 3: Addition and Subtraction

- Are students comfortable with combining and separating problems of the *unknown outcome* type? Do they know what they need to find in order to solve the problems? Can they keep the situation in mind as they solve them? Which students are comfortable with *combining with unknown change?*

- What strategies do students rely on for solving combining and separating problems? Do they count out all the quantities in the problem? Do they start with one quantity and count on or back? Do they use strategies involving number combinations? Are students taking apart numbers in ways that help them solve problems more easily? Can they keep track of the parts of the numbers they create and what to do with them? Which students are beginning to use strategies involving tens and ones?

- Can students record their strategies for solving combining and separating problems in a way that makes sense, using some combination of pictures, words, numbers, and equations?

In the *Investigations* curriculum, mathematical vocabulary is introduced naturally during the activities. We don't ask students to learn definitions of new terms; rather, they come to understand such words as *triangle, add, compare, data,* and *graph* by hearing them used frequently in discussion as they investigate new concepts. This approach is compatible with current theories of second-language acquisition, which emphasize the use of new vocabulary in meaningful contexts while students are actively involved with objects, pictures, and physical movement.

Listed below are some key words used in this unit that will not be new to most English speakers at this age level, but may be unfamiliar to students with limited English proficiency. You will want to spend additional time working on these words with your students who are learning English. If your students are working with a second-language teacher, you might enlist your colleague's aid in familiarizing students with these words, before and during this unit. In the classroom, look for opportunities for students to hear and use these words. Activities you can use to present the words are given in the appendix, Vocabulary Support for Second-Language Learners (p. 174).

red, blue, green These three colors are a key part of the Crayon Puzzles students solve in Investigation 1.

people, house, home In Investigation 2, while working with twos, students report on how many people live at home with them, and thus how many hands there are at home.

coins, cents, penny, nickel, dime, quarter For the game Collect 25¢ Together, students work with these U.S. coins (although some students may work almost exclusively with pennies).

imagine, story problem Students need to recognize the action in a story problem in order to determine what they need to find out. As the teacher describes a scenario for a story problem, students are asked to imagine what is happening.

Note: Many ESL programs have units on *colors* and *money.* Students would benefit from work on these before or during this mathematics unit.

About Story Problems

For Investigation 3, students do a lot of work with story problems, both as a class and individually during Choice Time (see Story Problems, Sets A–E, provided with the blackline masters at the end of this unit). These story problems are a very important part of the unit, and all students need a chance to work on them. To be sure the story problems are comprehensible to second-language learners, you may want to arrange for help from bilingual aides or parents to translate these problems, and if possible, ask them to be present in class during those sessions. If you do not have that support, take the time to make rebus drawings on the story problem cards as a comprehension aid.

Multicultural Extensions for All Students

- If you are creating your own story problems, include cultural references that are familiar to your students. For example, you might emphasize characteristic foods, clothing, games, or special events familiar to the students.
- Students will play card games involving numbers throughout this unit. They may know of other games involving numbers from their own or other cultures; if so, invite students to share these.
- After students play the game Collect 25¢ Together in Investigation 2, the class could explore the money used in another country. Students might bring in foreign coins to share.

Investigations

Number Combinations

What Happens

Session 1: Pictures of 10 In the Quick Images activity, students are briefly shown images, in this case arrangements of ten dots. After the image is removed, students make a copy of it, compare their copy with the original image, and share ways they "saw" the image. Students are introduced to using equations to describe the groups or rows of dots in each image.

Sessions 2 and 3: Number Combinations In these Choice Time sessions, students play two games: Dot Addition, in which they combine dot cards to make given numbers, and On and Off, in which they toss counters over a piece of paper and record how many land on and off the paper. At the end of Session 3, the whole class repeats Quick Images, this time with combinations of the Dot Addition Cards.

Sessions 4 and 5: Total of 10 Students learn a new game, Total of 10, in which they make combinations of 10 from a set of Number Cards. This game is added to Dot Addition and On and Off for Choice Time. At the end of Choice Time, students share some of the combinations they found while playing Dot Addition.

Session 6: How Many of Each Color? Students solve a How Many of Each? problem about 10 crayons in three different colors, finding one or more combinations of red, blue, and green crayons they could have to make 10 in all. They record and share their solutions. At the end of the session, the class repeats Quick Images with combinations of Dot Addition Cards.

Session 7, 8, and 9: Crayon Puzzles Students do Crayon Puzzles, which give a total number of red and blue crayons and a clue about how many of each color there is. They also play Counters in a Cup, in which one student secretly hides some counters, and a partner uses the remaining counters to determine how many have been hidden. These two activities and Total of 10 are offered during Choice Time. At the end of Choice Time, students share combinations of 10 they have been finding in the activities for this investigation.

Session 10 (Excursion): Number Combination Stories Students listen to the story *Ten Flashing Fireflies,* by Philemon Sturges, or another story that illustrates different combinations of a given number. Then they create their own number combination stories.

Routines Refer to the section About Classroom Routines (pp. 166–173) for suggestions on integrating into the school day regular practice of mathematical skills in counting, exploring data, and understanding time and changes.

Mathematical Emphasis

- Finding combinations of numbers up to about 20
- Finding the total of two or more single-digit numbers
- Exploring relationships among different combinations of a number
- Developing strategies for counting and combining dots arranged in rows or groups
- Reasoning about more, less, and equal amounts
- Using equations to describe arrangements of objects or pictures in groups

What to Plan Ahead of Time

Materials

- Overhead projector (Sessions 1–3, 6–7)
- Number Cards: 1 deck per pair, manufactured or make your own (Sessions 4–9)
- Empty crayon box or other box that holds ten crayons, with an opening large enough that everyone can see the colors of the crayons inside (Session 6)
- Red, blue, and green crayons: several of each color (Sessions 6–9)
- Cubes, counters, or crayons in three different colors, preferably red, blue, and green: at least 10 of each color per student (Sessions 6–9)
- Paper cups: 4–6 for the class (Sessions 7–9)
- *Ten Flashing Fireflies* by Philemon Sturges (Session 10, optional)
- Counters (such as buttons, bread tabs, or pennies): at least 30 per student (available)
- Chart paper or newsprint (18 by 24 inches): 15–20 sheets (available for use as needed)
- Unlined paper (available for student use)
- Envelopes for storing Dot Addition Card transparencies and Crayon Puzzles
- Resealable plastic bags to store cards

Other Preparation

- Before Session 6, separate your class set of interlocking cubes by color (perhaps with student help). Put each color in its own box or container.
- If you plan to provide folders in which students will save their work for the entire unit, prepare these for distribution.
- Duplicate the following student sheets and teaching resources, located at the end of this unit. If you have Student Activity Booklets, copy only items marked with an asterisk.

  For Session 1

  Family Letter* (p. 176): 1 per family (sign and date before copying)

  Quick Image Pictures of 10* (p. 186): 1 transparency. Cut apart the six images and store in an envelope. Note: Before making the transparencies, test an image on the overhead. If it seems too small, enlarge the images first on your copier or draw them larger on transparency film.

  For Sessions 2 and 3

  Student Sheet 1, Dot Addition (p. 177): 1 per student, homework

  Student Sheets 2–4, Dot Addition Boards A–C (pp. 178–180): 1 of each per pair and a few extras* (class), plus 1 each per student, homework

  Student Sheet 5, On and Off (p. 181): 1 per student, homework

  Student Sheet 6, On and Off Game Grid (p. 182): 1 per student and a few extras* (class), plus 1 per student, homework

Continued on next page

What to Plan Ahead of Time (cont.)

Dot Addition Cards (p. 187): 1 per pair for class (preferably on card stock), plus 1 per student, homework, and 1 transparency*. Cut apart and store in plastic bags or envelopes. **Note:** If you did the unit Building Number Sense, you should already have Dot Addition Cards, and your students may still have them at home.

Blank Dot Addition Board* (p. 188): copy as needed for making different boards (optional)

Sessions 4 and 5

Student Sheet 7, Total of 10 (p. 183): 1 per student, homework

Number Cards (pp. 225–228): 1 set per student, homework. **Note:** Students may already have sets at home from previous units. If you do not have manufactured cards, you will also need a class set for each pair, preferably on card stock. Cut apart each set and store in a plastic resealable bag. Remove wild cards for use in this investigation. Save sets for use in Investigation 3.

For Sessions 7, 8, and 9

Student Sheet 8, Counters in a Cup (p. 184): 1 per student, homework

Student Sheet 9, Counters in a Cup Game Grid (p. 185): 1 per student and a few extras* (class), plus 1 per student, homework

Crayon Puzzles 1–10 (pp. 189–190): 1 per student and 1 extra set*. Cut apart and sort into ten envelopes. Paste one copy of the puzzle on the envelope for identification. Crayon Puzzles 11–16* (p. 191) are optional challenges. Copy what you think you will need and sort into envelopes as described for puzzles 1–10.

Pictures of 10

What Happens

In the Quick Images activity, students are briefly shown images, in this case arrangements of ten dots. After the image is removed, students make a copy of it, compare their copy with the original image, and share ways they "saw" the image. Students are introduced to using equations to describe the groups or rows of dots in each image. Their work focuses on:

- developing strategies for counting and combining dots arranged in rows or in groups
- using equations to describe arrangements of dots (how many in each group or row and how many in all)
- becoming familiar with combinations of 10
- analyzing visual images
- describing position of and spatial relationships among objects

Materials

- Overhead projector
- Quick Image Pictures of 10 transparencies
- Family letter (1 per student)

Activity

Quick Image Pictures of 10

If your class has worked in the grade 1 units *Building Number Sense* or *Quilt Squares and Block Towns,* the students will be familiar with the Quick Images activity. As students repeat Quick Images throughout the year, they gain experience both in analyzing visual images and using number relationships to describe patterns.

Introducing or Reviewing Quick Images If your students have done Quick Images before, they will probably need only a quick review. Ask for volunteers to share what they remember about the activity. If this is their first experience with Quick Images, introduce it by name and ask for students' ideas about what an image is. After they have shared their ideas, explain the activity.

***Image* is another name for a picture. We'll be looking at some pictures, or images, of groups of dots. We call the activity "Quick" Images because you'll only get to see the picture for a short time. You have to look quickly when I show it. Then I'll cover it up, and you will try to make a copy of the picture you saw.**

Pictures of 10 Gather students in an area where they will be able to see the image projected from the overhead. Distribute counters and pencil and paper to each student, but explain that they will not use these until *after* you have flashed the first image and taken it away again. Keeping their hands in their laps while they view the image helps them concentrate.

Start with image A from the Pictures of 10. Show the image for about 5 seconds and then cover it. Observe students to decide whether you need to adjust the amount of time. If you show the image for too long, students will draw from the picture rather than their image of it; if you show it too briefly, they will not have time to form a mental image.

After you have hidden the image, students use counters or pencil and paper to recreate the image they saw.

When students are ready, show the dot image again for another 5 seconds. Again, students should not be drawing or moving counters around, but just studying the image while it is projected. Then remove the image and let students revise their arrangement of counters or change their drawing.

Finally, show the image again and leave it visible for further revision and for checking the number of dots shown.

While students are checking their work, make a quick sketch of the dot pattern on a piece of chart paper or the board. Then ask for volunteers to talk about how they thought about the dot pattern as they tried to remember it.

What helped you remember what the image looked like? What helped you make your own copy?

After each student gives an explanation, record his or her ideas by circling the dots that student saw grouped together. For example, if a student saw the pattern as "four dots on one end, four dots on the other end, and two left in the middle," you might record it this way:

Introduce the idea of using an equation to show this way of breaking the image into parts.

Tamika said she saw a group of four on one end, a group of four on the other end, and two in the middle. How does that help you think about the total number of dots? How could we write that as an equation?

Some students may be comfortable using equations to describe the way they thought about the images, but others may not be ready to do this without some support.

Gather and record several ways of breaking the image into parts. If possible, use markers or chalk of different colors to show the different ways. Otherwise, sketch the dot image again to demonstrate each new way of seeing it. The **Dialogue Box,** Pictures and Equations (p. 8), shows how one teacher helped students use equations to describe their individual ways of viewing the dot patterns.

Repeat the activity with Pictures of 10 images B, C, and D. If students are finding these images challenging, present E and F for more practice. Follow these steps each time:

1. Flash the image for about 5 seconds.
2. Give students time to draw or build what they saw.
3. Flash the image for another 5 seconds or so.
4. Allow time for students to revise or finish their first idea.
5. Show the image a final time. Ask students to describe what helped them remember the image.

Your students are likely to notice that all the transparencies in this set show 10 dots in all. If no one mentions it, encourage them by asking what they notice about the total number of dots in each picture.

Making Pictures of 10 If time permits at the end of the session, give students a few minutes to find their own way to arrange 10 (or 8 or 12) counters so it's easy to tell how many dots there are in all. To explain the task, quickly sketch an arrangement of 10 dots in two rows (like image A) and a random arrangement of 10 dots. Ask students which picture makes it easier to count the dots, and why. Students record their arrangements on unlined paper. Then they circle the groups of dots they see in the picture and write an equation to match, just as you did when discussing their ways of seeing the Quick Images. As students work, circulate to observe whether their equations match the way they have grouped the dots.

Session 1 Follow-Up

Family Connection Send home the signed family letter or the *Investigations* at Home booklet to introduce your work in this unit.

Pictures and Equations

During Quick Image Pictures of 10 (p. 5), students are sharing their thinking about image A. The teacher sketches the image on the board and helps them find ways to use equations to describe how they broke the image into parts.

What did you notice about the image? What helped you remember it?

Chanthou: I saw lines of dots.

How many?

Chanthou: Two.

Jamaar: There are five in each row.

[The teacher circles each row of five.] **So, two rows of five dots. How does that help us think about the total number of dots?**

Jamaar: You add them together.

What numbers would you use?

Nadia: Five plus five.

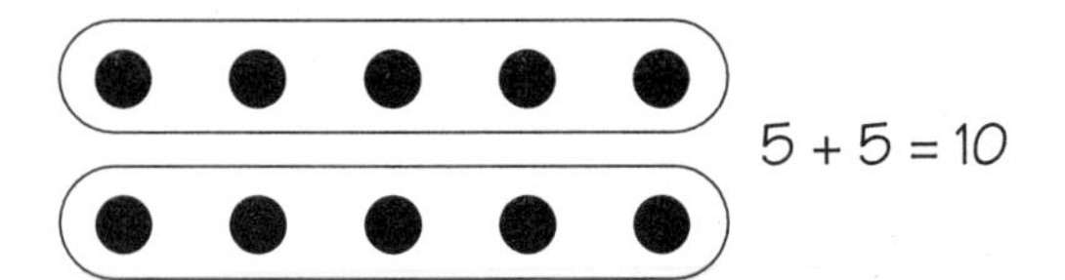

[The teacher records 5 + 5 = 10 to the right of the sketch.] **Five** *[points to the top row]* **and 5** *[the bottom row]* **is 10 in all. Who thought about this picture a different way?**

Diego: I saw twos. I counted by twos.

What did you say when you counted? Do you want to come up and show us?

[Diego comes to the overhead and counts each "column" of two: 2, 4, 6, 8, 10. The teacher circles each corresponding column on the board.]

How could you write it?

Diego: Two plus two is four. So, um... so, you write 2 + 2... That's four, so write + 4... then + 6?

[Several students raise their hands eagerly, but the teacher motions for silence.]

OK, let's take this one step at a time. What's this one *[points to the leftmost column]?*

Diego: Two.

[The teacher writes 2 above the column.] **And this next one** *[points to the next column]?*

Diego: Um... plus two.

[The teacher writes + 2.] **And?**

Diego: Two plus two... oh, plus two... plus two plus two. Plus two for each of those.

[The teacher writes in the remaining + 2's.] **OK, so, that's 2 + 2 + 2 + 2 + 2. And it equals?**

Max *[counting off each 2 on a finger]:* 2, 4, 6, 8, 10. Five sets of 2 equals 10.

2 + 2 + 2 + 2 + 2 = 10

[The teacher puts = 10 at the end of the expression.] **OK, five 2's equals 10. Is there another way?**

Tamika: One plus one plus one plus one plus one plus one plus... Wait. How many ones did I say?

How many ones are there in the picture? How many ones should there be?

Tamika: One for each dot. Ten.

So, what should we write?

[As Tamika counts, keeping track on her fingers, the teacher records each + 1 and then quickly draws a box around each dot.]

Let's count the 1's to make sure I have the right number of them. *[The class counts as the teacher points to each 1.]* **So, each of those 1's stands for one of the dots in the picture.**

1 + 1 + 1 + 1 + 1 + 1 + 1 + 1 + 1 + 1 = 10

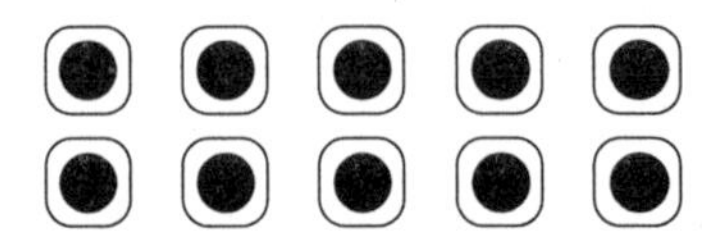

Number Combinations

What Happens

In these Choice Time sessions, students play two games: Dot Addition, in which they combine dot cards to make given numbers, and On and Off, in which they toss counters over a piece of paper and record how many land on and off the paper. At the end of Session 3, the whole class repeats Quick Images, this time with combinations of the Dot Addition Cards. Students' work focuses on:

- finding combinations of numbers up to about 20
- finding the total of several single-digit numbers
- developing strategies for counting and combining dots arranged in rows or in groups
- recording with addition notation

About Choice Time

Most of Sessions 2 and 3 will be Choice Time, a class structure used throughout the *Investigations* curriculum. See the **Teacher Note,** About Choice Time (p. 156), for information on how to set it up and how students can keep track of the choices they have completed.

The activity choices in Sessions 2 and 3 are Dot Addition and On and Off. Classes that have used the *Investigations* unit *Building Number Sense* will be familiar with both games. If that is the case, take just a few minutes at the start of Session 2 to review them before sending students to work on Choice Time.

If your students are new to both activities or have not done them in a while, you might spend all of Session 2 on Dot Addition, then at the start of Session 3 introduce (or review) the game On and Off, with Choice Time for the rest of that session.

Materials

- Dot Addition Cards (1 per student for class, and 1 per student, homework)
- Transparencies of Dot Addition Cards
- Student Sheet 1 (1 per student, homework)
- Student Sheets 2–4 (1 of each per pair plus extras for class, and 1 per student, homework)
- Blank Dot Addition Board (available as needed)
- Student Sheet 5 (1 per student, homework)
- Student Sheet 6 (1 per student plus extras for class, and 1 per student, homework)
- Unlined paper for game mats (1 sheet per pair, plus extras)
- Overhead projector

Activity

Dot Addition

Introduce or review the Dot Addition Cards with a round or two of Quick Images, using the transparencies of 4-dot and 5-dot cards. (See page 7 to review the steps of Quick Images.)

Then bring a set of Dot Addition Cards and Dot Addition Board A to the meeting area. Lay out three 2-dot cards, three 3's, three 4's, and two 5's on the floor so that everyone can see them. Mix the cards up, rather than grouping them by type.

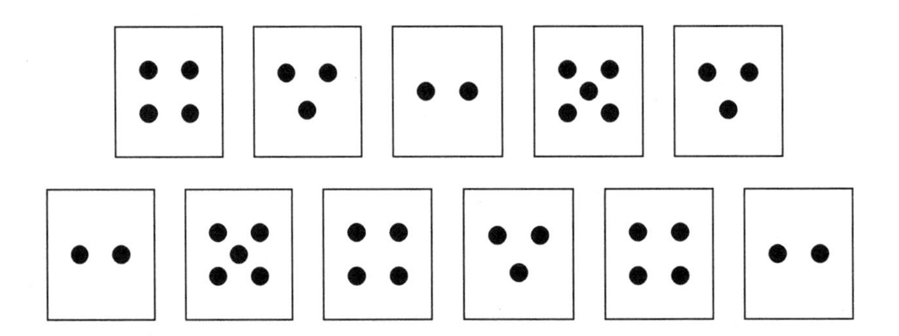

Who sees a way to put together some cards to make 9?... Michelle says 4 and 5. Can you show us how you know 4 and 5 make 9?... Michelle counted up from five: 6, 7, 8, 9. Does anyone have another way to show that 4 and 5 is 9?

Jonah says he knows 4 and 4 is 8, and 1 more is 9. Does anyone see a different way to use the cards to make 9?

Encourage students to find two or three different ways to show that each combination is correct. After students have suggested a few combinations of 9, ask them to find ways to make 12 with these cards.

Demonstrating the Game Set up a sample game. Explain that each pair gets a set of cards with 2 to 5 dots. At the start of the game, students lay out these cards, faceup, in rows of five. Demonstrate this layout. After you have put out two rows of cards and part of a third, pause and ask students to figure out how many cards are out. Ask for a few volunteers to share their thinking. Do they count each card? Do they count by 5's? Do they know the number combination 5 + 5?

When you have put out four rows of five cards, ask students to figure out how many cards are out and take a few moments for some students to share strategies.

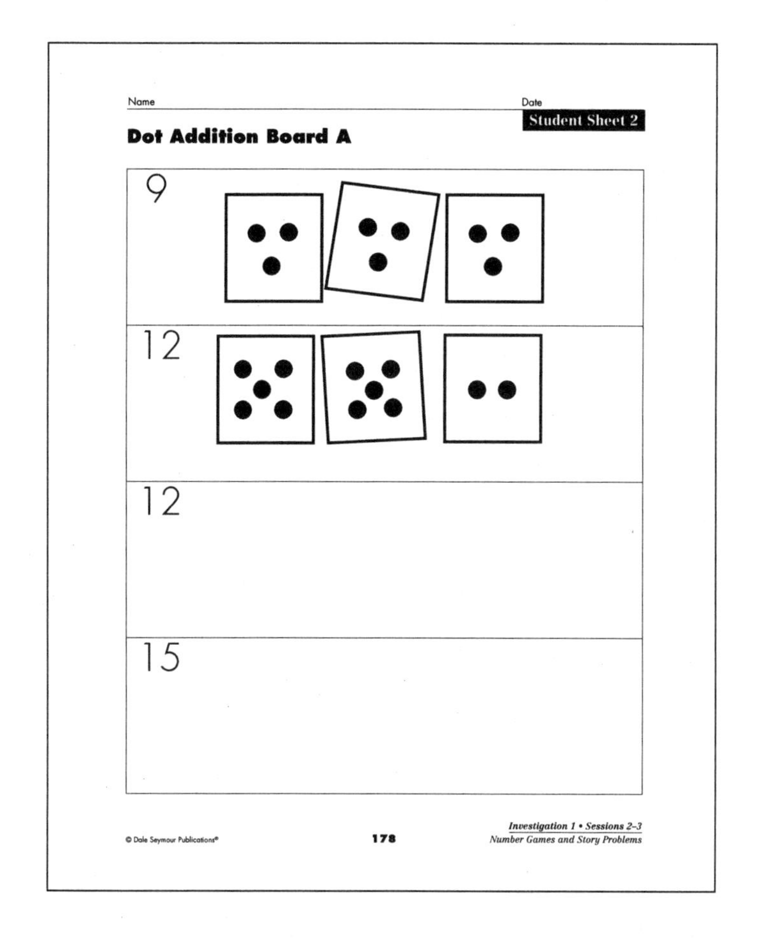

Name Date

Student Sheet 2

Dot Addition Board A

9
12
12
15

© Dale Seymour Publications®

178

Investigation 1 • Sessions 2-3
Number Games and Story Problems

Now lay Dot Addition Board A next to the array of cards. Explain that the object is to move cards onto the board to make a combination for each number on the board (9, 12, 12, and 15). Players can't use a card twice; so, for example, if they use three 3's to make 9, they won't be able to use four 3's to make 12. When a number appears twice on a board (as 12 does in this case), they need to find *two different ways* to make that number. Players can rearrange their cards at any time until they have a completed sheet.

As you demonstrate the game, involve students by asking for a volunteer to suggest a combination for each number. When you have filled the board, demonstrate recording each combination with addition notation on a separate piece of paper. For example, if three 3's were used to make 9, record it this way:

$$9 \quad 3 + 3 + 3$$

During Choice Time, students play Dot Addition in pairs, working with one set of cards and finding the combinations together. However, each student in the pair does his or her recording of the combinations on a separate sheet.

Activity

On and Off

Bring a sheet of unlined paper, a copy of the On and Off Game Grid, and 10 counters to the meeting area. (If working with a large class, you might quickly copy an enlarged game grid onto chart paper.) Place the sheet of paper down as a game mat and scatter the counters on it. Ask students how many counters there are. Then gather them up in one hand.

I have 10 counters. I'm going to toss them over this sheet of paper. What do you think might happen? Let's see how many land on the paper and how many land off the paper.

Toss the counters over the paper in such a way that some land on, some off the sheet. Ask how many landed *on* the paper and how many landed *off*. Then demonstrate filling in the game grid. Write 10 in the Total Number blank, then fill in the top row according to your first toss.

Repeat the activity until students understand the steps. Each time, record your results in the *On* and *Off* columns of the game grid.

During Choice Time, students play On and Off alone or in pairs.

Activity

Choice Time

Post a list of the two choices with a quick sketch as a visual reminder of the activity. Explain that for the rest of this math class and most of tomorrow's session, students will work on these two choices.

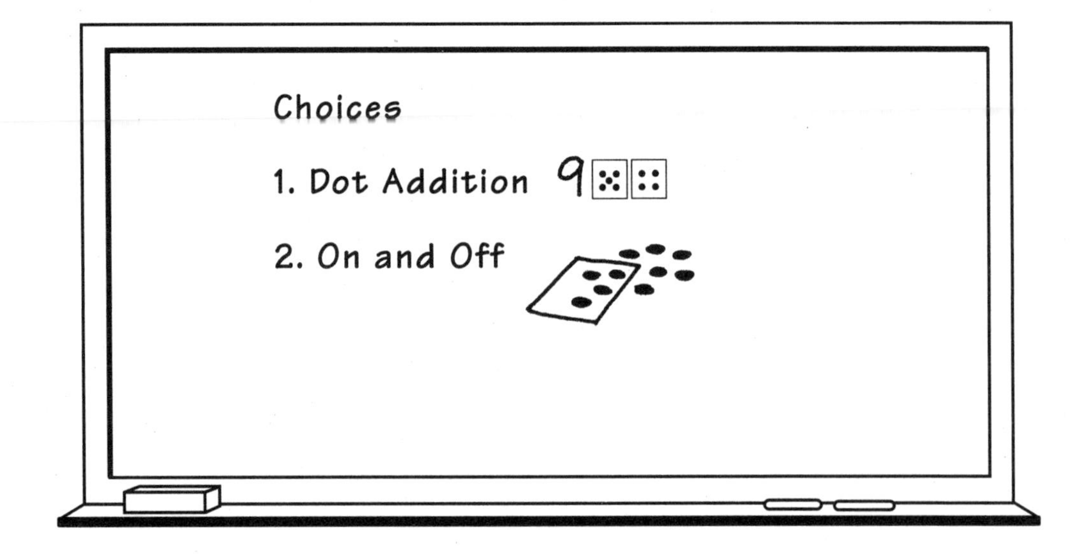

Students need to keep track of the choices they complete, and these records should be kept throughout an investigation, as the same choices may be offered in several different Choice Times. See the **Teacher Note,** About Choice Time (p. 156), for some suggestions on recordkeeping.

Choice 1: Dot Addition

Materials: Dot Addition Cards (1 set per pair); Student Sheets 2–4, Dot Addition Boards A–C (1 each per pair, plus extras); lined or unlined paper. Also have available copies of the Blank Dot Addition Board for making new boards with different numbers, as needed.

Pairs lay out their cards faceup in four rows of five and place one of the Dot Addition Boards beside them. They work together to find combinations of dots for each number on their board, using each card only once. When a number appears twice on a board, they need to find two different ways to make that number. When they finish, each student records on a separate sheet the addition combinations the pair made on their board.

Pairs may use the boards in any order. Many will have time to work with all three boards, especially if they play the game as homework. Some students may have time to do some boards twice. The second time, they look for different combinations for each number. If any students complete their work with all three boards and want to continue playing, use copies of the Blank Dot Addition Board to make new boards for them (for example, use the numbers 11, 13, 17, and 18, or 11, 11, 13, and 20).

Choice 2: On and Off

Materials: Counters (buttons, bread tabs); Student Sheet 6, On and Off Game Grid (1 per student, plus extras); unlined paper for game mats (1 per student or pair)

Students work alone or in pairs with a set of counters and a game mat. They toss the counters over the mat and record the number of counters that land *on* and *off* the paper, repeating the activity until they have completely filled a game grid. There is room to record two complete games on Student Sheet 6.

You might decide that all students work with the same number of counters, such as 12; you might let students choose their own total from among a range, such as 8 to 20; or you might assign particular totals to particular students, depending on the level of challenge you think they need. Whatever the case, be sure students record at the top of each game grid the total they are using for that game.

Observing the Students

Observe and listen to students as they work during Choice Time. Recording your observations will help you keep track of how students are interacting with materials and solving problems. The **Teacher Note,** Keeping Track of Students' Work (p. 158), offers some helpful strategies. Use the following questions to guide your observations of their work on Dot Addition and On and Off.

Dot Addition

- How do students construct their sums? Do they take a random approach, trying some cards, then discarding those and trying different cards, until they find a combination that works? Do they begin with one number, then count up to see what other number they need? Do they look for particular combinations? ("I want a 4 and a 5 to make 9, but I don't have a 5 left, so I could make 5 with a 2 and a 3.")

  If some students are finding the game very difficult, use the Blank Dot Addition Board to make them new boards with smaller numbers (such as 6, 6, 8, and 8, or 6, 7, 8, and 10).

- When counting to find the total of two or more cards, do students count out each quantity, from 1? Do they begin with the number of dots on one card and count up? Do any students count by numbers other than 1?

  Some students may begin counting by 2's or 3's for small quantities, and then switch to 1's as quantities increase. Understanding of counting by numbers other than 1 develops gradually over the early elementary years. Students will have opportunities to begin making sense of this in Investigation 2 and will continue to develop their understanding through the second and third grade *Investigations* curriculum.

- Do students use their knowledge of number combinations? What number combinations do they know? Do they know doubles (such as 3 + 3, or 4 + 4)? Can they quickly add 1 or 2 to another number?
- Are students comfortable recording with addition notation? If not, you might let them leave out the plus (+) sign and just record the individual numbers they used to make the totals.

If you can't tell how students are making their sums, ask them to explain their strategies. Ask, for example:

How did you figure out this combination for 10?

What if you couldn't use any 5's. Could you figure out something that would work for 10?

I see you've started with a 5-dot card to make 12. How are you figuring out what else you need?

On and Off

- Can students count the counters accurately? Do they organize the counters in some way to make them easier to count?
- Do students count both those on and those off the paper, or do they use one of the numbers to find the other? ("There are 8 on the paper, so there must be 4 off the paper, because 8 and 4 is 12.") Do they notice if the numbers they have recorded do not sum to the correct total, if, for example, they have counted incorrectly, or one of the counters is misplaced?
- Are students recording the numbers in the appropriate columns on the game grid? To check students' understanding of the relationship between the game grid and the way the counters landed, choose one of the number pairs on their grid and ask students to model it with counters on and off the paper.

Students ready for more challenge can think about the following: The game grid shows different ways to break the total number of counters into two parts. Can they think of a way to do this that is not on their game grid?

Activity

Quick Image Dot Combinations

About 20 minutes before the end of Session 3, announce the end of Choice Time. After cleanup of the Choice Time materials, call students together for two or three rounds of Quick Images, using transparencies of the Dot Addition Cards in various combinations. (See page 7 to review the steps of Quick Images.)

Try repeating the same number card two, three, or four times in a row. Here are four possibilities, for example:

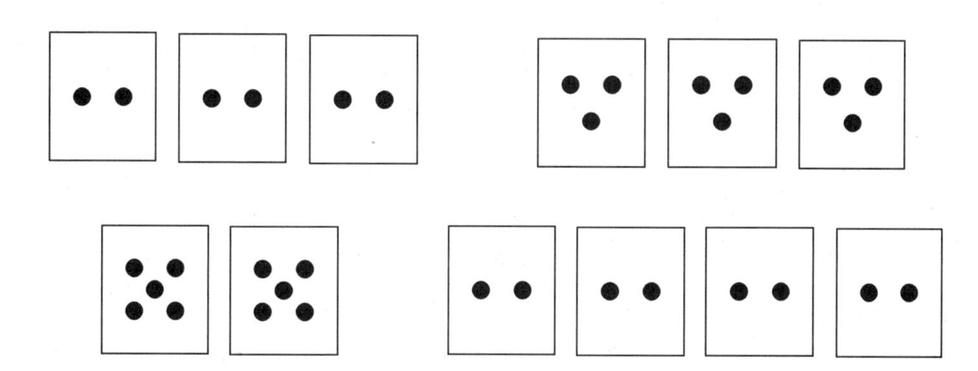

Sessions 2 and 3 Follow-Up

Homework

Dot Addition Students play Dot Addition with someone at home. They may take home Dot Addition Boards they have not yet used in class, or they may use the same boards again, this time making each number in a different way. They will also need Student Sheet 1, Dot Addition, and a set of Dot Addition Cards, although if you have presented the unit *Building Number Sense,* they may already have both the directions and a set of cards at home.

Some teachers provide an envelope or a "Math at Home" folder for storing the materials (such as Dot Addition Cards) students bring home throughout the year. In any case, suggest that students keep their cards and game materials in a special place at home because they will be playing math games for homework throughout the unit.

On and Off While Dot Addition offers good homework for these sessions, you may also want to send home the game On and Off for use now or later in the unit. Students will need a copy of Student Sheet 5, On and Off, and Student Sheet 6, On and Off Game Grid. They will also need counters such as buttons, bread tabs, paper clips, or pennies. You might suggest a total number for students to use, or give them a range of numbers to choose from (perhaps 12 to 20). If students are choosing their own number, ask them to select one they have not yet used in class for this game.

Extensions

Most and Fewest Using any of the Dot Addition Boards A–C, students explore the following questions:

> How can you make each number using the *fewest* Dot Addition Cards? Is there more than one way?
>
> How can you make each number using the *most* Dot Addition Cards?

Combinations of One Number Challenge students to see which numbers, up to 20, they can make with repetitions of a single Dot Addition Card. For example, they might make 6 with two 3's or with three 2's. They will need to pool two sets of cards to make some numbers. Be sure they understand that they will not be able to make every number; for example, they cannot make 5 with just one type of Dot Addition Card.

Total of 10

What Happens

Students learn a new game, Total of 10, in which they make combinations of 10 from a set of Number Cards. This game is added to Dot Addition and On and Off for Choice Time. At the end of Choice Time, students share some of the combinations they found while playing Dot Addition. Their work focuses on:

- finding combinations of numbers up to about 20
- finding the total of two or more single-digit numbers
- exploring relationships among different combinations of a number
- developing strategies for counting and combining dots arranged in rows or in groups
- recording with addition notation

Materials

- Number Cards (1 deck per pair, and 1 deck per student, homework)
- Student Sheet 7 (1 per student, homework)
- Counters (available)
- Dot Addition Cards, Dot Addition Boards A–C, Blank Dot Addition Boards, and On and Off Game Grids (from the previous Choice Time)

Activity

Introducing Total of 10

Total of 10 is played with a deck of Number Cards. To introduce the game, gather students around you and play a demonstration game with a student volunteer.

I'm going to show you how to play a game called Total of 10. To begin this game, we mix up a deck of Number Cards, then lay them out faceup in rows of five. We'll make four rows, with five cards in each row. *[Demonstrate.]*

Explain that the object of the game is to find combinations of two or more cards that total 10. As you play the demonstration game, involve students in your turn.

I'll start. Does anyone see some cards I could put together to make 10?

If students need help thinking of possible combinations, scan the cards for a pair that totals 10 and point to one of the cards in the pair.

What if I want to use this 7? Is there a card I could put together with this 7 to make 10?

Remove the cards that make 10, and set them aside in a pile.

Next, the student volunteer takes a turn. Again, offer hints if the player does not readily find a combination of 10.

Let's say William wants to use this 9. Is there another card he could put together with the 9 to make 10?

If no one suggests making 10 with more than two cards, do this on your next turn. Scan the remaining cards to find a set of three or more that total 10. Point to two of the cards, and ask students what other card (or cards) would go with those to make 10. Put your second combination of 10 in a separate pile, so it doesn't get mixed up with your first.

Continue playing until no more combinations of 10 can be made. Explain that players should try to use as many of the cards as possible. Sometimes there will be several cards left, but occasionally players may be able to use them all. A 0 card may be included in any combination, and a 10 card by itself is one way to make 10.

At the end of the game, players turn over each of their piles and list the combinations of 10 they made, using addition notation. Model this for the class.

Tell students that the rest of Sessions 4 and 5 will be Choice Time, and the game Total of 10 is one of the choices.

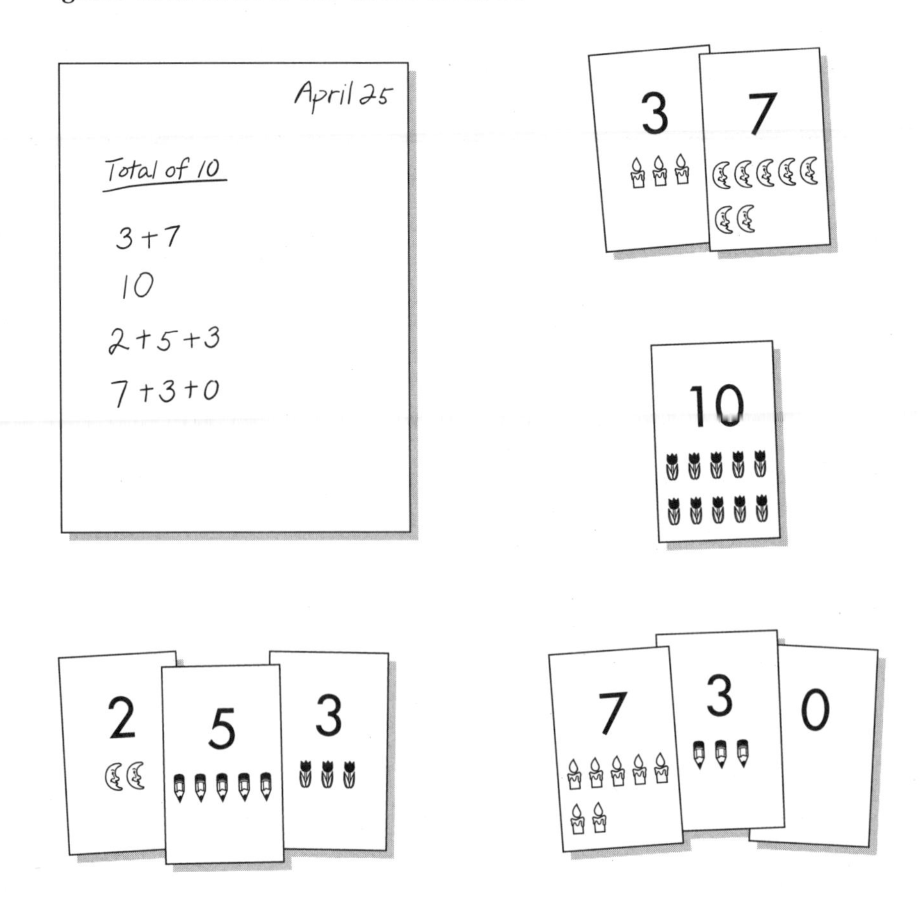

Choice Time

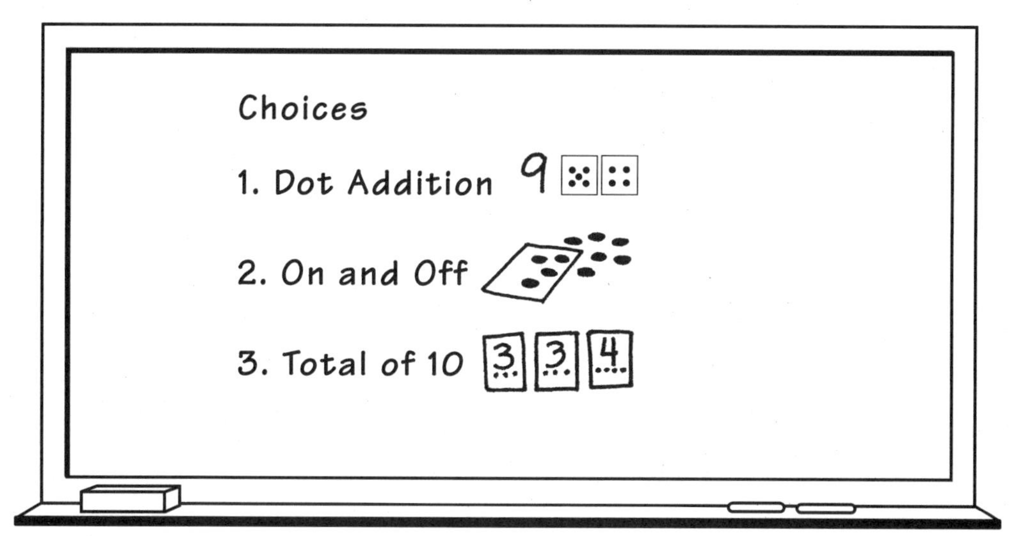

Post a list of the choices and call attention to the three options. Advise students that by the end of this Choice Time, they must have completed their work on Dot Addition.

See p. 13 to review Choices 1 and 2, Dot Addition and On and Off.

Choice 3: Total of 10

Materials: Number Cards (1 deck per pair, with wild cards removed); lined or unlined paper; counters (available)

Students play in pairs; if necessary, three can play together, or a student can play alone. To start, players lay out 20 cards faceup in four rows of five. They put the other cards aside. Players take turns finding a combination of two or more cards that totals 10 exactly. A player removes these cards from the layout and sets them aside, placing each combination in a separate pile so they don't get mixed up until the combinations can be recorded at the end of the game. Players try to use as many cards as possible. The game is over when no more combinations of 10 can be made. Each player then lists on a separate sheet the combinations of 10 that he or she made.

Students ready for more challenge can try one of the following variations:

- Play with wild cards, which may be used as any number.
- Each time cards are removed from the layout, replace them with new cards from the deck.
- Play to make combinations of a larger number, such as 20. For this variation, players will need to replace each card used with new cards from the deck. This version is introduced to the whole class in Investigation 3.

Observing the Students

See pp. 14–15 for guidelines on observing students' work on Dot Addition and On and Off.

Total of 10

- What strategies do students use to find their combinations? Do they seem to work randomly, choosing a number to start with, and then combining it with different numbers until they find a way to make 10 in all? ("There's a 3. This card is a 5. Would that work? Let's see... 4, 5, 6, 7, 8. I need more. What if I also added this 3?") Do they keep track of how many more they need to make 10? ("I have a 3 and a 2, that's 5, so I need 5 more, because 5 and 5 is 10.") Do they look for particular combinations of 10? ("Is there an 8 and a 2 left?")

  If some students find it overwhelming to choose combinations of 10 from among 20 cards, you might suggest they look only for pairs that total 10.
- How do students combine numbers? Do they count from 1 each time? Do they count up from one of the numbers? Do they use knowledge of number combinations? Do they use number combinations they know to find new combinations?
- How do students determine that the game is over? Do they keep trying combinations of remaining cards to make 10? Do they reason about the cards that remain? ("There's a 5, 8, and 7 left. I know we're done, because the two smallest ones are 5 and 7, and 5 and 5 is 10, so it's more.")
- Do they record the combinations accurately? Do they use addition notation correctly?
- Do students play cooperatively and help one another with the game? If you think some students are playing too competitively, remind them that the goal is to work together to use as many cards as possible from the layout. The best players are not those who use more cards or make more combinations, but the ones who play cooperatively, by checking or helping one another, by explaining their thinking to one another, by asking a partner for help, or by waiting while a partner takes the time to look for a combination of 10.

Activity

Sharing Dot Addition Sums

Near the end of Session 5, call students together to share some of the combinations they found when playing Dot Addition. Students will need the papers on which they recorded their work for this activity.

Choose one of the numbers, such as 12, for which students found combinations.

What's one way you made 12 with your cards?... Nathan got 3 plus 3 plus 3 plus 3. Did anyone else get that? Can you show how you know that makes 12?... So, Nathan knew three 3's is 9, 1 more is 10, and 2 more than 10 is 12. Does anyone have a different way to show that 3 plus 3 plus 3 plus 3 is 12?... Who made 12 a different way?

Ask for two or three different ways to show that each combination is correct. Some students may build from combinations they already know; others may not yet reason this way, still seeing each combination as a completely separate problem. Students will increasingly recognize relationships among combinations as they continue through the early elementary grades. The **Teacher Note,** Strategies for Learning Addition Combinations (p. 159), offers more information on learning addition combinations (number combinations with two addends) in the early elementary grades. See also the **Teacher Note,** Building on Number Combinations You Know (p. 22), for ways that students may begin to use familiar combinations to help them figure out unfamiliar ones as they work on the activities in this investigation.

Record each combination students suggest. As the list grows, before you record a new combination, ask students to check first if is already listed. By asking them to attend to the solutions already recorded, you help them appreciate the importance of keeping track. This may also help some students to notice relationships among different solutions.

Students may wonder whether combinations that use the same addends in different order (such as 4 + 3 + 3 and 3 + 4 + 3) are "the same." You might let the class decide whether or not to include such combinations on the list. If students suggest combinations of 12 that can't be made with Dot Addition Cards (such as 2 + 2 + 8), you might decide to record those as well, perhaps in a different list.

When students have listed all the combinations of 12 they made, repeat the activity for one or two other totals students worked with, such as 10 or 20.

Sessions 4 and 5 Follow-Up

Total of 10 Students play Total of 10 with someone at home. They will need directions (Student Sheet 7) and a deck of Number Cards. (If you have presented other *Investigations* units, students may already have Number Cards at home.) If you want homework turned in, ask students to bring back to class their lists of combinations of 10 they make for each game they play.

Teacher Note

Building on Number Combinations You Know

In first grade, most students are just becoming familiar with the addition combinations from 0 + 0 to 10 + 10. (See the **Teacher Note,** Strategies for Learning Addition Combinations, p. 159.) As they work with numbers in many different ways, some first graders begin to reason about the combinations they already know to figure out others. During this investigation, you may observe some of the following ways of thinking.

Counting On from a Known Combination When playing Dot Addition (p. 13), one student explained that she made 8 with 3 and 5 because "I know 3 and 3 is 6, and 2 more is 7, 8. But I didn't have enough 3's, so I put 2 and 3 together and made it 5." This approach is similar to *counting on,* a strategy that many students use for finding the total of two numbers. That is, to combine two numbers, they start at one and count up the other: "7 and 3 is — 8, 9, 10." Here, instead of beginning with a number, the student began with a number combination: 3 + 3. She counted up 2 from the total (6) to reach the goal number (8), and then increased one of the addends in the combination accordingly (3 + 5).

Counting Back from a Known Combination One student used a counting back approach when he was playing Counters in a Cup (p. 36) with 15 counters. His partner had hidden some of the 15 counters in a cup, and there were 6 remaining. This student reasoned that 9 were hidden because "10 plus 6 is 16, so take away 1 from 10 and it's 9, so 9 plus 6 is 15."

Probably your students will use counting back to find a new number combination only occasionally, and only with very familiar combinations. Counting backwards and building numbers by "taking away" is less familiar to first graders than counting up and building numbers by combining parts.

Breaking a Familiar Combination into Parts and Recombining the Parts Another way that students find new combinations was demonstrated by a girl playing the game Total of 10 (p. 19). Here's her explanation of how she found the combination 6, 3, and 1: "Last turn, I did 7 and 2 and 1. So this time I took a 6, and I knew 6 and 1 is 7, and so it's like 6 and 1, and 2, and 1. So 3, because 1 plus 2. And then 1 left."

Another student made 15 in Dot Addition by beginning with a combination he knew, 8 + 7. To use the Dot Addition Cards, he needed numbers 5 or less, so he broke 8 into 5 and 3, and 7 into 5 and 2: 5 + 3 + 5 + 2. He checked his work by regrouping the addends so that he could use combinations he knew: 5 and 5 is 10, 3 and 2 is 5, and 10 and 5 is 15.

In both of these examples, the student recognized that total remains the same, regardless of how the addends are broken into parts and regardless of the order in which the addends are combined. This flexible way of thinking about numbers is very powerful. When the second student combined the two 5's to make 10 and the 2 and 3 to make 5, he decided how to group the addends in the problem to make it easier to solve; he did not simply combine addends in the order in which they appeared. Note that he also used 10 as a familiar "landing place" in his calculation. He found a way to combine the addends to make 10, and then constructed a sum in which 10 is an addend. In the next year or two, as students gain more experience with two-digit numbers, they will continue to develop strategies involving tens and ones.

Adjusting the Numbers in a Familiar Combination Some first graders begin to recognize that if they *add* an amount to one addend in a number combination and take away the same amount from another, the total remains the same.

In the Crayon Puzzles activity (p. 35), one of the Challenge puzzles states: "I have 11 [red and blue] crayons.... I have one more blue than red. How many of each could I have?" One student explained that he approached this puzzle by recording the first combination of 11 that came to mind, 4 + 7. He figured out that these two numbers were 3 apart and decided to make them closer together by adding 1 to the 4 and taking away 1 from the 7. He knew the total wouldn't change because "it's just adding 1 here and taking it away there, it's still 11."

To help your students think about building from number combinations they know, offer time for them to share strategies they have used in playing the games in this investigation. However, avoid directly teaching these strategies. Although some students may begin this kind of reasoning with problems that involve smaller, more familiar numbers, they probably need many more experiences with number combinations over the next year or two before they will be able to use such strategies more broadly.

Session 6

How Many of Each Color?

Materials

- Red, blue, and green crayons (several of each color)
- Empty box that holds ten crayons
- Cubes, counters, crayons, or other materials in three colors, preferably red, blue, and green (10 of each color per student)
- Lined or unlined paper
- Transparencies of Dot Addition Cards
- Overhead projector

What Happens

Students solve a How Many of Each? problem about 10 crayons in three different colors, finding one or more combinations of red, blue, and green crayons they could have to make 10 in all. They record and share their solutions. At the end of the session, the class repeats Quick Images with combinations of Dot Addition Cards. Students' work focuses on:

- finding combinations of 10
- recording solutions with pictures, numbers, and words
- finding more than one solution to a problem

Activity

Teacher Checkpoint

Ten Crayons

Introducing the Activity Ten Crayons is what we term in the *Investigations* curriculum a How Many of Each? problem, for which students find combinations of two or more things that make up a given total. These problems give students practice with number combinations, with developing strategies for combining quantities, and with recording and organizing their solutions. Classes that have done the units *Mathematical Thinking at Grade 1* or *Building Number Sense* will be familiar with this problem type and the activity will need only a brief introduction.

For this introduction, have several red, blue, and green crayons and an empty box that holds at least ten crayons.

I want to put 10 crayons in my box. I need some red ones, some blue ones, and some green ones. What could I have? How many reds? How many blues? How many greens? Remember, I need 10 crayons in all.

Accept two different suggestions and model them by counting out crayons and putting them in the box. If some students disagree, ask them to explain their thinking, but keep the discussion brief and keep the focus on explaining the task clearly enough so that students can find solutions on their own. Explain that there is more than one correct solution to this problem, and that students are to find solutions different from the ones just suggested.

Solving the Problem Students may work alone, in pairs, or in small groups. Encourage them to share ideas with one another. Make available cubes, crayons, or other materials in three colors for those who want to use them. When students find a solution, they record it on paper using pictures, numbers, words, or a combination of these. Each solution should include the number of each color of crayon. Encourage them to find more than one solution.

2 green + 4 red
1 2 3 4 5 6
4 blue
7 8 9 10

red Green blue
2 + 4 + 4
red Green blue
3 + 2 + 5
red blue Green
8 + 1 + 1
Green blue red
7 + 1 + 2
blue red Green
6 + 2 + 2

Observing the Students

This Teacher Checkpoint gives you the opportunity to observe students' strategies for finding combinations of 10 with three addends. As students are working, circulate to observe how they are approaching the problem and to offer support as needed.

- Do students understand what they are to find? Do they have a sense of how to begin?

  In order to solve this problem, students need to keep in mind the number of each color and the total number of crayons. If some students have difficulty keeping track of all this information, ask them first to find a combination of 10 crayons with just two colors:

  Let's say you have ten crayons in all, some of them are red, and some of them are blue. How many of each could you have?

  When they have found a solution using crayons or counters, ask them to replace some of the blues with greens.

  What if you wanted some green crayons, too? Could you trade some greens for some blues? Now, how many of each color do you have? Do you still have 10 in all?

  As you offer assistance, be careful not to show students a solution, as this can make it difficult for them to find their own solutions.

- What strategies do students have for solving the problem? Do they collect and count objects in three colors and adjust the number until they have 10? Do they use number combinations? ("5 and 5 is 10, and 2 and 3 is 5, so 2, 3, and 5 would work." Or, "4 red and 3 blue is 7 in all, so I need 8, 9, 10—3 greens.")

- How do students record their solutions? Do they use pictures? numbers? words? equations? If some students record only the number of each color (for example, "2 and 3 and 5"), ask them to find a way to show what each number represents.

  How can we tell if your 2 means two blues, two reds, or two greens?

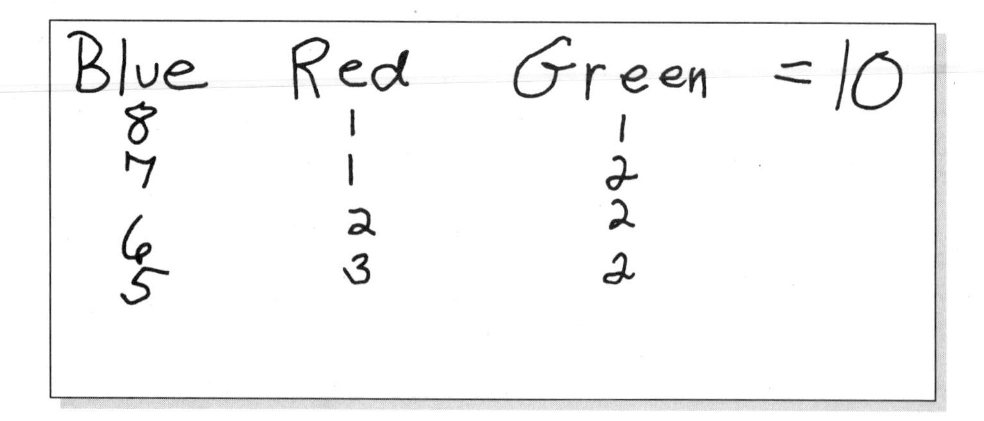

- Do students find more than one solution? How do they do so? Do they take counters randomly until they find another solution? Or do they find new solutions by changing something about a solution they already have? ("I have 6 blues, 3 reds, 1 green. Take 1 blue and make it a red, and it's 5 blues, 4 reds, 1 green.") Do they compare new solutions with those already on their list? Do they notice that some solutions are the same as others with the numbers rearranged?

  Some students are sufficiently challenged by finding just one solution; you can help them appreciate the variety of solutions by asking them to share their work with a partner. Other students may want to find all the possible solutions, especially if they successfully found all the solutions to How Many of Each? problems with only two different things earlier in the year. Encourage these students to persevere, but let them know that problems involving three different things (like three colors) have many more solutions.

- Do students check that their solutions are correct? What strategies do they have for checking?

4 Blue's 3green's and 3 red's

4+3+3

3+3=6
4+6=10

Activity

Sharing Solutions

When most students have found several solutions, call the class together to share some of their work.

Who has a solution to share?... Leah got 3 red, 4 blue, and 3 green. Did anyone else get that? Can you show how you know that makes 10?... Does anyone have a different way to show that 3, 4, and 3 make 10?

Record each solution on a piece of chart paper or on the board with the corresponding equation.

10 Crayons			
Red	Blue	Green	
3	4	3	3 + 4 + 3 = 10
8	1	1	8 + 1 + 1 = 10
7	1	2	7 + 1 + 2 = 10
4	3	3	4 + 3 + 3 = 10

For each new solution suggested, ask the class to check if it is the same as any you have already recorded. Some students may wonder if combinations that use the same numbers in a different order are "the same." Help these students think about what the columns on the chart show.

Someone said 4, 3, and 3 isn't on the list yet, but several of you point out that we've already listed 3, 4, and 3. What do the numbers in this left column show? Yes, the number of blue crayons. The numbers in the middle? On the right? So, we have 3 reds, 4 blues, and 3 greens. Do we have one with 4 reds, 3 blues, and 3 greens?

After several minutes, some students may find it difficult to remain engaged and focused, perhaps because the particular solutions they found are already listed. Before ending the discussion, acknowledge that there are more solutions. For those students still eager to share, you might set up a time later in the day when they can add their solutions to the chart with a partner or in a smaller group. Some students may want to continue finding more solutions in their free time during the day or at home.

Activity

Quick Image Dot Combinations

Spend about 15 minutes at the end of Session 6 on Quick Images. (See p. 7 to review the steps of Quick Images.) You will probably have time to present two or three images, using three or four Dot Addition Cards in combinations:

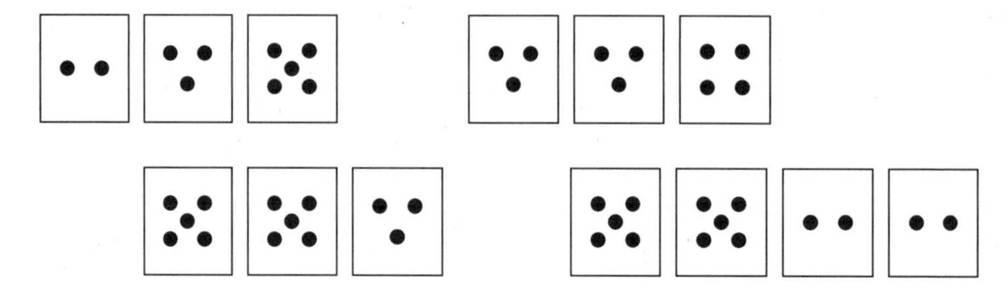

Session 6 Follow-Up

Combining Solutions to Make 20 Pairs of students combine their solutions to the Ten Crayons problem. They find out how many of each color they have, and then how many crayons they have in all. For example: One student has 5 red, 4 blue, and 1 green. The other has 7 red, 1 blue, and 2 green. Together they have a total of 12 red, 5 blue, and 3 green, and 20 crayons in all.

Fifteen Crayons Students solve the same problem with a different total of crayons, such as 15 or 25.

Crayons in Four Colors Students find combinations of *four* different colors that make 10 crayons in all.

Ways to Make 10 Students find and record different ways to make 10. They may use calculators. For example:

$3 + 1 + 1 + 1 + 4 = 10$
$20 - 10 = 10$

Sessions 7, 8, and 9

Crayon Puzzles

Materials

- Cubes, counters, crayons, or other materials in three colors, preferably red, blue, and green (10 of each color per student)
- Crayon Puzzles 1–10 (in prepared envelopes)
- Crayon Puzzles 11–16 (in prepared envelopes, optional challenges)
- Unlined paper
- Paste or glue sticks
- Number Cards (1 deck per pair)
- Student Sheet 8 (1 per student, homework)
- Student Sheet 9 (1 per student and extras for class, plus 1 per student, homework)
- Paper cups (4–6 for the class)
- Counters (buttons, bread tabls)

What Happens

Students do Crayon Puzzles, which give a total number of red and blue crayons and a clue about how many of each color there is. They also play Counters in a Cup, in which one student secretly hides some counters, and a partner uses the remaining counters to determine how many have been hidden. These two activities and Total of 10 are offered during Choice Time. At the end of Choice Time, students share combinations of 10 they have been finding in the activities for this investigation. Their work focuses on:

- finding combinations of numbers up to about 20
- exploring relationships among different combinations of a number
- reasoning about more, less, and equal amounts
- finding the total of two or more single-digit numbers

CRAYON PUZZLE 1

I have 6 crayons.
Some are blue and some are red.
I have the same number of each color.
How many of each could I have?

CRAYON
CRAYON
CRAYON
CRAYON
CRAYON
CRAYON

Activity

Introducing Crayon Puzzles

Crayon Puzzle 1 Take the envelope with copies of Crayon Puzzle 1 and distribute one to each student. Make available crayons, counters, cubes, or other materials in two colors (if possible, red and blue), and paper for recording solutions. After students paste or glue their puzzle to a blank sheet, read the puzzle aloud with the class (see above).

Students work on the puzzle alone or in pairs. Some may find it helpful to model the problem with counters, pictures, or numbers. They record their solutions using pictures, numbers, words, or a combination of these, being sure to specify how many of each color.

First grade students differ widely in how readily they can solve this puzzle. Although similar to How Many of Each? problems, the Crayon Puzzles are more challenging because they contain another clue—the relative size of each set—which introduces an additional condition students' answers must meet. Circulate quickly as students begin working to see who is ready to move on to another puzzle and who needs more help.

For Students Having Difficulty Ask students to explain the puzzle in their own words. Sometimes in talking through the puzzle, they will see a way to approach it. If any students are still uncertain of how to begin, model the strategy of starting with any set of crayons that matches the total and then adjusting it. That is, put out six crayons or other counters, taking more of one color than the other (for example, one red crayon and five blue crayons). Ask students if there are the same number of red and blue, and then ask them to adjust the set until that they have the same number of each.

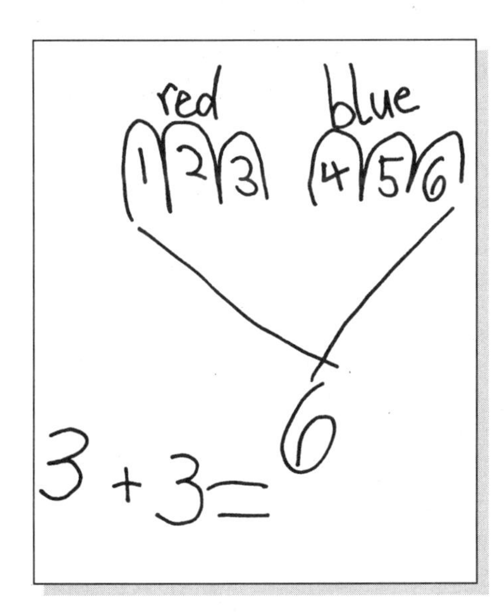

Crayon Puzzle 2 Distribute the second Crayon Puzzle to individuals or small groups as they complete the first puzzle. They may glue this second puzzle to the same or another sheet. Read it aloud with them.

> I have 7 crayons. Some are blue and some are red. I have more blue crayons. How many of each could I have?

To make sure that students understand the clue involving *more*, ask them to put the puzzle into their own words. If some students need help making sense of the clue, put out seven interlocking cubes using more reds than blues, and ask which color there are *more* of. Then, ask students to replace some of the reds with blues until there are more blues. Some students may find it helpful to connect the cubes of each color in stacks; they can then compare the heights of the stacks to find which has more.

CRAYON PUZZLE 2

I have 7 crayons.
Some are blue and some
are red.
I have more blue crayons.
How many of each could
I have?

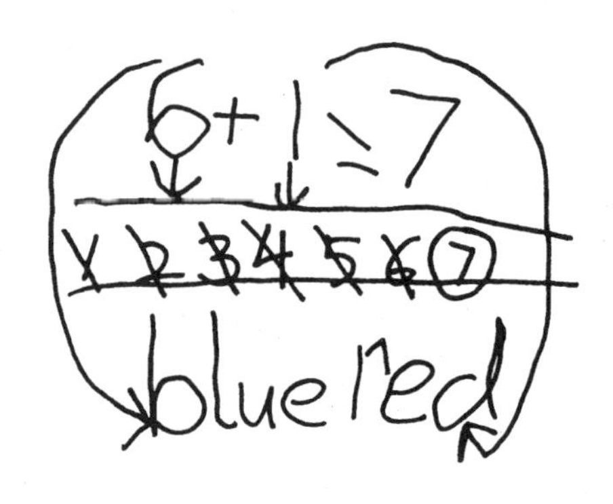

Blue
Red
I have 5 Blue's
and I have 2 Red's
and I got up 7
I have 6 Blue's
and 1 Red

Again, students may vary widely in how readily they make sense of and solve the puzzle. Some may adjust their solution to the first puzzle. ("In that one, 3 and 3 is 6, so 4 blue and 3 red make 7.") Others may use their knowledge of number combinations; and still others may make different combinations of seven with red and blue counters until they find a combination in which there are more blues.

Some students have difficulty keeping in mind all the puzzle clues. When they have found a way to make seven with blue and red crayons, you may need to remind them to check if they have more blues than reds.

Students who finish early could try to find more than one solution. As needed, make available Crayon Puzzles 3–10 for them to work on until everyone has finished Puzzles 1 and 2.

Activity

Counters in a Cup

Note: For Choice Time during the rest of Sessions 7, 8, and 9, the activity choices are Total of 10, Crayon Puzzles, and a new activity, Counters in a Cup. Classes that have done the unit *Building Number Sense* will be familiar with Counters in a Cup; after a brief whole-class review of this game, they can proceed with Choice Time. If your class does not know this game, you might decide to offer only the other two choices during Session 7, then introduce Counters in a Cup more extensively at the start of Session 8.

To introduce this game, put eight counters and an upside-down paper cup in front of you.

How many counters do I have? Yes, I have eight counters and an empty cup. *[Hold up the cup so everyone can see that it is empty.]* **I'm going to hide a secret number of counters under this cup.**

Gather up the counters and, in such a way that students cannot see what you are doing, hide three of them under the overturned cup. Then place the remaining counters in front of the cup so students can see them.

Think silently to yourself about how many I hid. When you think you know, raise your hand, but please don't say anything.... OK, how many counters did I hide? How do you know?

Garret thinks three are hidden, because he counted the five counters and then counted up three more—six, seven, eight. Who has another idea about how many are hidden? How do you know?

Keep asking for other ideas about the number of counters hidden, even after someone gives the correct answer. That way, you help students think for themselves about whether the solution has been found.

When the solution has been checked by lifting the cup, demonstrate filling in the appropriate information on Student Sheet 9, Counters in a Cup Game Grid. Start by writing 8 in the Total Number blank for the first game, then record the number *inside* and the number *outside* the cup in the appropriate column in the first row.

Repeat the game once or twice, until you think students understand the steps. Vary the number of counters that you hide. For each round, show how to record the results on the game grid.

Explain that during Choice Time, students will play this game with a partner. One student will hide a secret number of the counters under a cup, and the other will figure out the number hidden; then they will switch roles.

Activity

Choice Time

On your Choices list, cross off Dot Addition and On and Off, and add the two new activities (or only Crayon Puzzles, if you plan to introduce Counters in a Cup in Session 8). Explain that by the end of three sessions, students should have tried all the choices. Remind them of your class system for keeping track of the choices they have done.

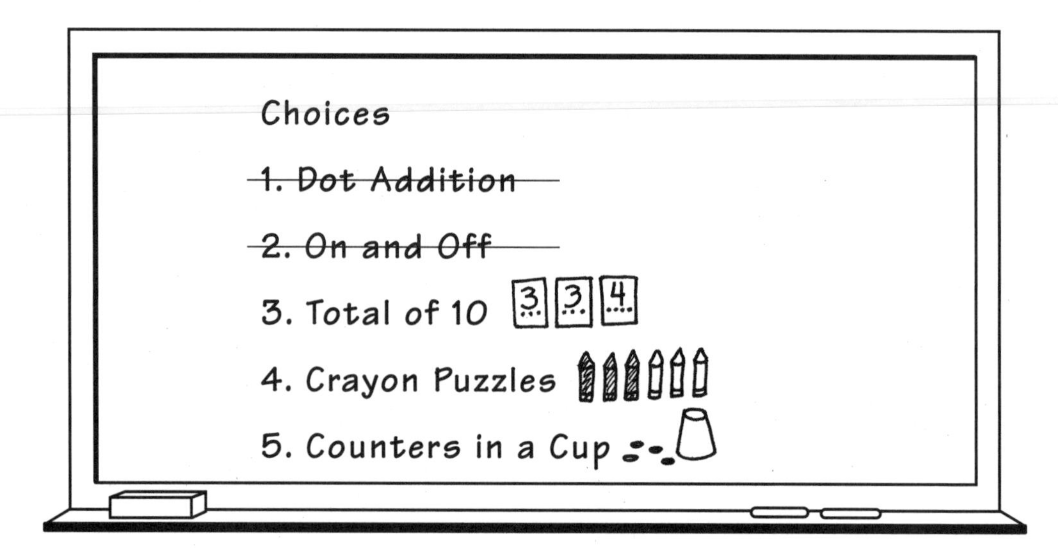

For a review of Choice 3: Total of 10, see p. 19.

Choice 4: Crayon Puzzles

Materials: Envelopes with copies of Crayon Puzzles 3–10; crayons, cubes, counters, or other materials in three different colors (preferably red, blue, and green); unlined paper; paste or glue sticks. (Have puzzles 11–16 available for students ready for more challenge.)

Students continue solving Crayon Puzzles, working alone or in pairs. They paste or glue each puzzle to a sheet of paper on which they record their solution with pictures, numbers, or words, being sure to specify how many crayons they have of each color.

Students may work on the puzzles in any order, although the puzzles generally increase in difficulty. As needed, set up procedures for getting help in reading the puzzles. Certain students may be designated as reading helpers for their peers, or you may want to start off a group of students yourself by reading the puzzle with them.

❖ **Tip for the Linguistically Diverse Classroom** Read each Crayon Puzzle aloud as students make their own rebus drawings over key words. For example, they may use crayon marks in the actual colors for red, blue, and green, circle the color there is more of, and so forth.

While some students may not have time to complete all the Crayon Puzzles during Choice Time, others may complete them quickly and be ready for more challenge. These students may work on Crayon Puzzles 11–16. Puzzle 16 includes the clue "I have fewer red crayons." You might check that students can make sense of *fewer*, and help them model the situation as needed.

☆☆☆☆☆ CRAYON PUZZLE 16☆

I have 15 crayons.
Some are blue and some are red.
I have fewer red crayons.
How many of each could I have?

red	blue
7	8
6	9
5	10
4	11
3	12
2	13
1	14

Choice 5: Counters in a Cup

Materials: Paper cups (1 per pair); counters (buttons, bread tabs); Student Sheet 9, Counters in a Cup Game Grid (1 per pair, plus extras)

Each pair has a set of counters and a cup. You might decide that all students work with a particular total number of counters, such as 10; you might let students choose a total of counters from among a range, such as 7 to 15; or you might assign particular totals to particular pairs, depending on the level of challenge you think they need. Players record the total number of counters they are working with at the top of the game grid.

To start, one player hides a secret number of the counters under the cup, and the other then figures out the number hidden. After each turn, players record on the game grid the number of counters both in and out of the cup. Players take turns, repeating the game until they have completely filled the grid. Encourage players to vary the number of counters they hide, although they may occasionally hide the same number.

As you observe students working, you may want to suggest they work with more or fewer counters. The game can be challenging for some students, as it involves keeping track of and coordinating three amounts: the total number of counters, the number visible, and the number hidden. Students who are having difficulty could use a smaller total, such as 5 counters.

Some students may need to do this activity several times with a total of 5 or 6 counters before they can work with a set as large as 8 or 10. Others may be ready to work with an even larger total, such as 12 or 15. Once you think students are working at an appropriate level of challenge, be sure they have a chance to play the game several times with the same total.

Observing the Students

To review the guidelines for observing students at work on Choice 3: Total of 10, see p. 20.

Crayon Puzzles

- Can students make sense of the puzzles? Can they keep all the clues in mind? Read through the clues with any students having difficulty, and ask them to explain the puzzle in their own words. Some students may need reminders that their solutions must match *all* the clues. You might encourage the strategy of first finding a combination that matches the total and then adjusting it to match the clue about the relative number of reds and blues.

- What strategies do students have for solving the puzzles? Do they seem to generate combinations randomly until they find one that matches all the clues? Do they use their knowledge of number combinations? Do they build upon their solutions to other Crayon Puzzles? ("I did 6 blues and 3 reds for more blues than reds, so I can do 3 blues and 6 reds for more reds than blues.") Do they find a set of blue and red crayons or counters that matches the total, and then adjust until they find a solution with the appropriate amounts of each color?
- Do students look for more than one solution? Do they treat each new solution as a separate problem? Do they find new solutions by changing something about a solution they already have?

Students who finish early can try to find all possible solutions to each puzzle. They can also make up their own Crayon Puzzles to share with a partner.

Counters in a Cup

- What strategies do students have for finding the number of counters hidden? Do they seem to guess at how many are hidden? Do they count all the counters they see, and then count on to the total number of counters? Do they use knowledge of number combinations? Do they use relationships among combinations? ("Last time, there were 3 outside the cup and 7 hidden. This time, there are 4 outside, so there must be 6 hidden—because there's 1 more out, so there has to be 1 less in the cup.")
- How do students go about varying the number of counters they hide? Do they seem to randomly choose a number? Do they look at the game grid to find a number of counters that has not yet been hidden? Do they have a systematic approach, such as first hiding 1 counter, then 2, and so on? (If students do not think of hiding all or none, do not suggest this; they enjoy discovering this "trick" on their own or learning it from classmates.)
- Do students understand how to use the game grid? Do they know how to use 0 to show when *all* the counters are either hidden or visible?

Activity

Sharing Combinations of 10

Near the end of Session 9, call students together to share a few combinations of 10. They may refer to their work for Total of 10, Ten Crayons, or any other activities in which they found combinations of 10. Students might also come up with combinations of 10 they haven't previously found. As necessary, remind the class that you are looking for ways to make 10 by adding numbers together, not by subtracting.

Record each combination on the board or chart paper. Before recording each suggestion, ask students to check if it is already listed. This can sometimes help students begin to notice relationships among solutions. See the **Dialogue Box**, Combinations of 10 (p. 39), for an example of how this discussion went in one class.

Before ending the discussion, acknowledge that there are other combinations of 10. For further work on this, see the Extension activity, Finding All the Combinations of 10.

Sessions 7, 8, and 9 Follow-Up

Counters in a Cup Students teach someone at home to play Counters in a Cup. Send home a copy of Student Sheet 8, Counters in a Cup, and Student Sheet 9, Counters in a Cup Game Grid. They will need a paper cup (or other small, opaque container) and something to use as counters (such as buttons, bread tabs, paper clips, or pennies). Either suggest a total number of counters for students to use, or ask them to pick their own total from a range, such as 10 to 15. If students are choosing a number, remind them to select one they have not yet used in class for Counters in a Cup.

Math Games At any time for homework, ask students to play at home one of the other games introduced in this unit. For example, they might choose from Total of 10 and On and Off. Sometimes they might try a new or more challenging variation of a game.

Finding All the Combinations of 10 Students consider their class list of combinations of 10 and think about whether it contains all possible combinations. You may want to leave your list posted for another day or two and suggest that students who think they have found other combinations of 10 add them to the list.

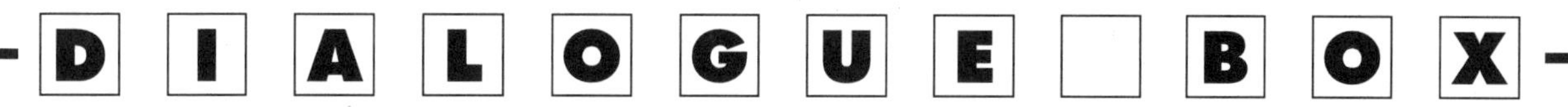

Combinations of 10

After Choice Time in Sessions 7–9, this class is sharing combinations of 10 they have been finding (p. 37). The teacher records each suggestion on chart paper. As students compare the combinations in the growing list, they begin to explore relationships among them.

Who can give us a combination of 10?

Claire: 8 + 1 + 1.

How do you know that's 10?

Claire: *[Holds up 8 fingers]* 8 and 1 *[another finger]* is 9, and 1 more *[another finger]* is 10.

Who has another combination of 10?

William: 3 and 7. I thought of 7, and I counted up 8, 9, 10.

Jonah: 5 + 5.

How do you know?

Jonah: Because 5 *[holds up one hand]* and 5 *[holds up the other hand]*—10 in all.

Tuan: I have another one: 5 + 4 + 1.

How do you know that's 10?

Tuan: 4 + 1 is 5. And it's like what Jonah did—5 and 5.

Yukiko: You can do it to both!

Do what to both?

Yukiko: Both 5's. They can both be 4 + 1.

So, you can break up each 5 into 4 and 1, to get 4 + 1 + 4 + 1. Who has something else?

Luis: 3 + 1 + 6.

Shavonne: We already have it, in a way.

Can you tell us what you mean?

Shavonne: It's like 3 + 7, because 3 and 1 and 6... You put the 1 in the 6. Then it's the same.

Shavonne is saying they're similar— if you have 3 + 1 + 6 and you add up the 1 and the 6, you end up with 3 + 7.

Kristi Ann: Here's one: 4 and 4 and 1 and 1.

How do you know that's 10?

Kristi Ann: I took the first one [the first expression, 8 + 1 + 1] and made it 4 and 4.

[When the teacher adds 4 + 4 + 1 + 1 to the list, students call out that it's already there.]

We have 4 + 4 + 1 + 1, and 4 + 1 + 4 + 1. What do you think? Are they the same?

Max: They're the same numbers.

Kristi Ann: My way is in a different order.

The same numbers in a different order. They both have two 4's and two 1's. Yukiko thought of 5 and 5, and she broke each 5 into 4 and 1. Kristi Ann thought of it as 8 and 1 and 1, and broke the 8 into 4 and 4.

Iris: I have a new way: 8 and 2.

Claire: We already have that. It's the first one.

[Pointing to 8 + 1 + 1] **I don't see a 2.**

Claire: But 1 and 1 is 2. It's the same thing.

Iris: I put the 1 + 1 together, so it looks like 8 + 1 + 1, but it isn't. They used three numbers to make 10, and I used two numbers because I put the 1 and 1 together.

Session 10 (Excursion)

Number Combination Stories

Materials

- *Ten Flashing Fireflies* by Philemon Sturges (or similar book)
- Drawing paper
- Crayons or markers

What Happens

Students listen to the story *Ten Flashing Fireflies,* by Philemon Sturges, or another story that illustrates different combinations of a given number. Then they create their own number combination stories. Their work focuses on:

- finding combinations of a given number
- showing number combinations with pictures, words, and equations

Activity

Reading a Number Combination Book

Each facing two pages of *Ten Flashing Fireflies* by Philemon Sturges shows a different combination of 10 with two addends: some of the 10 fireflies are inside a jar, and some are outside a jar.

If you cannot get this book, look for a different book about number combinations. For example, try one of these:

- *Anno's Counting House* by Mitsumasa Anno (Philomel Books, 1982) illustrates combinations of ten with two addends. This picture book (no words) follows the progress of ten people moving their belongings from their old house to a new house.
- *Ten for Dinner* by Jo Ellen Bogart (Scholastic, 1989) illustrates combinations of ten with several addends. In this story, ten children come to a birthday party, with varying groups arriving at different times and in different clothes, choosing different foods and different things to do.
- *Six Sleepy Sheep* by Jeffie Ross Gordon (Boyds Mills Press, Caroline House, 1991) illustrates combinations of six with two or more addends. At bedtime, while some of the six sheep drop off to sleep, others can't sleep and keep busy doing other things.

The following activity can be easily adapted for any of these stories.

Read *Ten Flashing Fireflies* aloud and show the pictures. Stop two or three times while reading to cover the fireflies on one page with your hand or with a blank sheet of paper, and ask students to make predictions about the number of fireflies on the other page. For example, cover the page on which six fireflies are flying free and ask:

How many fireflies are in the jar? Yes, there are four. How many do you think are on the page I've covered? How do you know?... Nadia thinks there are six on that page because she counted on her fingers up from four— 5, 6, 7, 8, 9, 10. Does anyone else predict that there are six? Why do you think so?... Does anyone predict something different?

When you have finished reading, ask students to list the combinations of 10 in the story with you. Make a two-column chart on chart paper or the board, with one column headed *IN* and the other *OUT.*

The story showed a lot of different ways that the 10 fireflies could be in and out of the jar. Can you remember any of them? How many were in the jar? How many were out of the jar?

Record combinations in the order in which students suggest them, and write the corresponding equations:

IN	OUT	
0	10	$0 + 10 = 10$
5	5	$5 + 5 = 10$
1	9	$1 + 9 = 10$
9	1	$9 + 1 = 10$
4	6	$4 + 6 = 10$

For each new combination, ask the class to check if it is already listed. If students wonder if the order of the numbers matters (that is, is 4 and 6 the same as 6 and 4?), ask them to think about what the columns on the chart tell us.

When students have listed all the combinations they can think of, flip through the book with them to see if they have recorded all the combinations in the book. On each spread of facing pages, students count the number in and out of the jar with you, then look for that combination on their list. Record any additional combinations and the corresponding equations.

Is there a way of making 10 that the book didn't show? How do you know?

If no one notices that the combinations you have listed only involve two addends, bring up this idea yourself.

Think of some of the ways you've made 10 in the last couple of weeks. Did you do anything that's not on this list? What about when you solved the problem about 10 blue, red, and green crayons?

Accept a few suggestions of ways to make 10 with more than two addends and add the corresponding equations to the class list. If necessary, students can look back through their math folders for their work on Ten Crayons, Pictures of 10 Dots, Dot Addition, or Total of 10.

To make these new equations meaningful in the context of the fireflies story, ask students to think about what the *third* addend might indicate.

What if the first number shows the number of fireflies in a jar, and the second number shows the number of fireflies flying around in the sky. What could this third number show? What could a third group of fireflies be doing?

Encourage students to brainstorm a few ideas. For example, the third group might be in a tree, in someone's hand, in the house, or sitting on the dog. If students are having difficulty coming up with possibilities, hold up one or more of the pictures that show fireflies in a variety of places: six fireflies in a jar, three fireflies in the sky, and one in a hat; two fireflies in a jar, seven in the sky, and one in the girl's hand; and seven in the jar, one high in the sky, and two on the ground. Students can then brainstorm other places, realistic or silly, that fireflies could be.

Repeat the process with another equation with three or more addends. Take a couple of minutes for a few students to share their ideas about a situation this equation could represent.

Activity

Creating Number Combination Stories

As a follow-up activity, each student creates a one-page number combination "story." This is the task: Students find a combination of a given number; they illustrate it with a picture of animals or objects in different groups; they write an equation to match; and they describe what is happening in words.

You will need to decide what number or numbers the combinations will be based on. You might decide that everyone works with combinations of the same number, such as 12, or students might choose from several numbers, such as 7, 10, 11, 12, and 15.

You may also want to choose a general theme. For some students, coming up with an idea for a story takes quite a bit of time. Choosing a class theme beforehand can help them focus instead on number combinations. You might suggest a context that everyone will use, such as animals on a farm or in a zoo, creatures on imaginary planets, or dinosaurs in a prehistoric jungle. Or you might brainstorm ideas with the class and ask them to vote on the possibilities generated, or allow them to choose any of the listed ideas.

Telling a little story can help set a context and serve as a model for what students will do. For example, with a farm context and a total of 12, you might tell a story like this:

Last Saturday I went to a farm and saw a lot of different animals. First I saw 12 pigs. Three pigs were rolling in the mud, seven pigs were sleeping, and two pigs were eating corn.

Draw a quick sketch to match your story. Then write the corresponding equation: 3 + 7 + 2 = 12. Ask students how your equation reflects what the animals are doing in the picture.

What does this 7 stand for? Which number stands for the pigs that are rolling in the mud?

You can also involve the whole group in creating a suitable situation:

I also saw 12 cats. What do you think some of them might have been doing?... Fernando thinks some of them might have been climbing trees. So, let's sketch that... How many cats should I put up in this tree?... OK, 5. What might another group of cats be doing?

When you have accounted for all 12 cats in your story and sketches, ask students for the matching equation and write it above your pictures.

Explain that for the rest of this session, students will make their own one-page story about the farm (or another theme), with a picture and equation to match, as you have just done. They also write (or dictate) a sentence about what the animals are doing. Give them the total number (or a choice of numbers) they are to work with, and explain that they decide for themselves how to break up the total into groups and what each group is doing. If they choose something that they find too hard to draw, they can draw just one or two and then write how many are in each group and what the groups are doing. Students who do not finish may complete their work outside of math class.

❖ **Tip for the Linguistically Diverse Classroom** Students may rely on their pictures and the equation to tell their story.

Observing the Students

- How do students come up with combinations of their total number? If some are having difficulty, encourage them to look for combinations that use only two addends. They can use counters to help them find combinations.
- Do their equations reflect the arrangements of objects they have drawn?

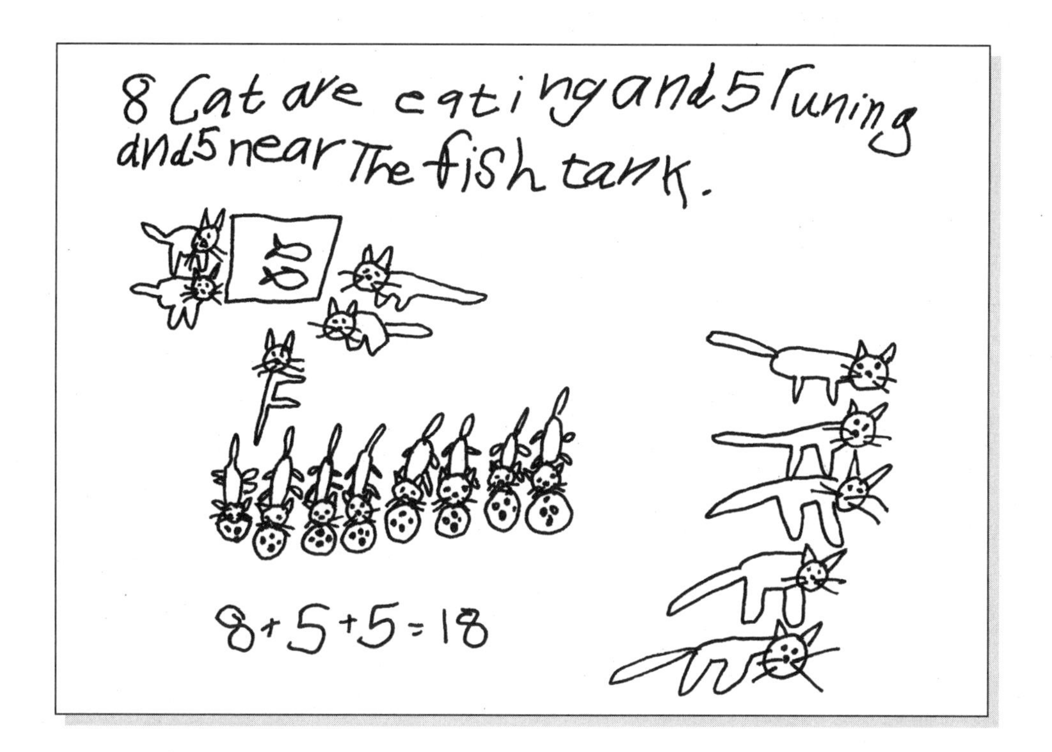

At the end of the session, reserve 5–10 minutes for a few students to share their work with the class. Ask students to explain what their page shows, what parts they broke the total into, and how they showed their visual arrangement with an equation.

You might post students' work on a board and encourage students to look at the posted pictures over the next day or so.

Session 10 Follow-Up

All the Combinations Suggest that students find *all* the combinations (with two addends) of the total they used for their story. They might also illustrate some or all of these with additional one-page stories, and collect them into a booklet.

INVESTIGATION 2

Twos, Fives, and Tens

What Happens

Session 1: How Many Hands? Students begin their work on finding the total of several 2's as they find how many hands a group of students has. They record their solution strategies and then share them with the class. For homework, they find the total number of hands at home.

Session 2: Twos and Fours The class lists and makes predictions about the number of hands for successive numbers of people. Then students find the total of several 4's in a problem about wheels on cars. They record and share their solution strategies.

Session 3: Collect 25¢ Together Students briefly explore coins, then play the game Collect 25¢ Together, rolling a dot cube and collecting coins until they have a total of 25¢.

Sessions 4 and 5: Counting and Combining Students do Quick Images with pictures of squares arranged in groups. In How Many Squares? they find the total number of squares in two sets of squares grouped by ones, twos, and fives. During Choice Time, they continue work on How Many Squares?, Collect 25¢ Together, and solving problems with 2's, 4's, or 5's. At the end of Choice Time, they share strategies for solving one of these problems.

Sessions 6, 7, and 8: Numbers to 100 The class plays Missing Numbers, identifying numbers that are missing from a 100 chart. This activity is then added to Choice Time, which continues over the three sessions. At the end of Choice Time, the class repeats Quick Images with squares and shares strategies for How Many Squares?

Session 9: Patterns of Fives and Tens Class starts with Missing Numbers. Then, after doing Clapping Patterns, students represent a five-part pattern on a 100 chart and in another way (perhaps with drawings, cubes, or pattern blocks).

Sessions 10, 11, and 12: Twos, Fives, and Tens The game Roll Tens, in which players collect interlocking cubes and group them in rows of ten, is added to Choice Time, along with Exploring Calculators. A variation on How Many Squares? is used as an assessment.

Session 13 (Excursion): Counting by Kangaroos Students listen to the story *Counting by Kangaroos,* by Joy N. Hulme, which illustrates finding the total of several equal amounts. Students then find the total for three groups of 11, 12, or 13 animals.

Routines Refer to the section About Classroom Routines (pp. 166–173) for suggestions on integrating into the school day regular practice of mathematical skills in counting, exploring data, and understanding time and changes.

Mathematical Emphasis

- Developing strategies for organizing sets of objects so that they are easy to count and combine
- Finding the total of several 2's, 4's, 5's, or 10's
- Recording strategies for counting and combining, using pictures, numbers, and words
- Reading, writing, and sequencing numbers to 100
- Becoming familiar with coins and equivalencies among them
- Working with different models for grouping 2's, 4's, 5's, and 10's
- Beginning to develop meaning for counting by 2's
- Exploring patterns in the number sequence
- Developing a sense of the size of the numbers up to 100
- Exploring calculators as a mathematical tool

What to Plan Ahead of Time

Materials

- Dot cubes: 1–2 per pair (Sessions 3–8, 10–12)
- Play coin sets: about 30 pennies, 6–7 nickels, 3–4 dimes, and 1–2 quarters per pair, stored in resealable plastic bags, paper cups, or other small containers (Sessions 3–8)
- Interlocking cubes: class set of 1000 (Sessions 4–12)
- Overhead projector (Sessions 4–8)
- Hundred Number Wall Chart with transparent pockets, numeral cards, and colored plastic chart markers (Sessions 6–9)
- Plastic Hundred Number Boards with removable tiles: 2 or 3 for the class (Sessions 6–9)
- Class sets of pattern blocks and other available manipulatives, for making patterns (Session 9)
- Calculators: at least 6–8 for the class (Sessions 10–12)
- Unlined paper (available for students, all sessions)
- Counters, such as buttons, bread tabs, or pennies: at least 40 per pair (available for use as needed)
- Chart paper or newsprint (18 by 24 inches): 15–20 sheets (available for use as needed)
- Envelopes for storing overhead transparencies and sets of Squares

Other Preparation

- To introduce coins in Session 3, provide play or real coins, at least one penny, nickel, dime, and quarter per student.
- Duplicate the following student sheets and teaching resources, located at the end of this unit. If you have Student Activity Booklets, copy only items marked with an asterisk.

For Session 1

Student Sheet 10, How Many Hands at Home? (p. 192): 1 per student, homework

For Session 2

Student Sheet 11, Cats and Paws (p. 193): 1 per student, homework

For Session 3

Student Sheet 12, Collect 25¢ Together (p. 194): 1 per student, homework

For Sessions 4 and 5

Quick Image Squares* (pp. 202–204): 1 transparency of each sheet. Cut apart and store in three envelopes. Save extra transparent squares from the first sheet to introduce the Squares activity.

Squares* (pp. 205–207): Duplicate on card stock or heavy colored paper (use a single color for all squares). Copy 2 sheets of singles, 3 sheets of pairs, and 3 sheets of five-square strips. Cut apart. See next page for assembly of Sets A, B, and C.

Student Sheet 13, Feet, Fingers, and Legs (pp. 195–196): 1 per student

Continued on next page

What to Plan Ahead of Time (cont.)

For Sessions 6, 7, and 8

Student Sheet 14, Coins (p. 197): 1 per student, homework

Session 9

Student Sheet 15, 100 Chart (p. 198): 1 per student and a few extras*

200 Chart* (p. 209): several for the class (optional, for challenge)

For Sessions 10, 11, and 12

Roll Tens Game Mats* (pp. 210–211): 5 of each, and extras available. Copy on colored paper. Cut apart each 30 mat and 50 mat. **Note:** These mats are designed for use with cubes ¾ inch on a side. If your cubes are a different size, draw mats to fit your cubes, based on these models.

Student Sheets 16–18, What's Missing? (pp. 199–201): 1 of each per student, homework

Blank 100 Chart* (p. 208): several for the class (optional)

- Before Sessions 4–5, assemble Squares, Sets A and B as follows:

  Set A: 8 singles and 6 pairs
  Set B: 7 singles and 4 fives

  Make four of each set. Store in separate envelopes labeled Set A and Set B.

- When students have finished with Squares, sets A and B, assemble Set C for Sessions 10–12.

  Set C: 5 singles, 5 pairs, and 2 fives

  Make about 12 sets for students to share. You can recombine the squares from Sets A and B, and cut fives into pairs as needed. Store in envelopes labeled Set C.

Session 1

How Many Hands?

What Happens

Students begin their work on finding the total of several 2's as they find how many hands a group of students has. They record their solution strategies and then share them with the class. For homework, they find the total number of hands at home. Students' work focuses on:

- finding the total of several 2's
- recording strategies for counting and combining using pictures, numbers, and words
- developing meaning for counting by 2's

Materials

- Unlined paper
- Student Sheet 10 (1 per student, homework)

Activity

Hands and Other Pairs

Begin by taking a few minutes for students to brainstorm things that come in pairs (2's). Record students' ideas in words and pictures on the board or chart paper. If no one suggests parts of the body, ask students to think of body parts that come in pairs (eyes, ears, feet, hands, and so forth). Note that this activity might evoke some sensitive issues; see the **Teacher Note,** Dealing with Sensitive Issues (p. 55), for more information.

Next, call a group of four students to the front of the room.

One thing that comes in pairs is hands. Suppose we wanted to find out how many hands there are in this group of four people. How could we figure that out?

Gather several different strategies. Although the total number of hands will surely be mentioned, keep the focus on solution strategies. The **Dialogue Box,** Four Students, Eight Hands (p. 56), gives an example of this discussion.

If the students at the front of the room get restless during the discussion, quickly sketch a stick figure to represent each student so they may return to their seats.

Activity

How Many Hands?

Students now work on their own to solve a similar problem, finding the total number of hands in a group of *eight* students. If many students found the previous problem (number of hands in a group of four) challenging, use a group of five or six students instead of eight. (You want to keep the numbers in this activity at a comfortable level, since finding a way to record solution strategies is itself a challenge.)

As you present the problem, emphasize that students are not just to find the answer, but also show how they found their solutions.

I'd like you to find the number of hands in a group of eight students. You can solve this problem in whatever way makes sense to you. You can use cubes, you can use your fingers, or you can use anything else you need. Keep track of how you solved the problem, and write or draw on paper how you solved it so that someone else can understand exactly what you did. You can use any combination of words, pictures, and numbers. When you have finished, we'll share some of your strategies.

Writing about how they solved a problem is challenging for young students. Although students who have completed the unit *Building Number Sense* will have had some experience recording their strategies, they need many experiences with this before they are comfortable recording their thinking. Over the next few weeks, concentrate on helping them record in a way that shows how they arrived at their solutions. See the **Teacher Note,** Writing and Recording (p. 161), for more on the importance and challenge of learning to record solution strategies.
The **Dialogue Box,** Helping Students Record Their Strategies (p. 57), demonstrates the difficulties one class had recording their thinking and ways that the teacher helped them.

Students may work alone or in pairs, but each student must record his or her own solution strategy.

Observing the Students

Circulate as students work to observe the following:

- How do students approach the problem? Do they model it with objects or pictures? Do they work mentally or count to themselves? Do they record numbers or equations to help them find the solution? If some students are finding the problem difficult, suggest they model it with objects or pictures. You might decide to ask some students to find the total number of hands in a smaller group.

- What strategies do students use? Do they use strategies based on counting by 1's? by 2's? Do they use doubles? ("Two people have 4 hands, so 4 people have 8 hands, and 8 have 16 hands.") Do they use strategies based on number combinations? ("I know 2 and 2 is 4, and 4 and 2 is 6, and 6 and 2 is 8..." and so on.) Do they break the problem into smaller parts and then combine the parts? ("I know 4 people have 8 hands, and 4 plus 4 is 8, so 8 people have 8 plus 8 hands.")

  Many students may use strategies that involve counting by 1's to find the total. Some may start out using strategies that involve 2's, doubles, or number combinations, but then return to counting by 1's as the numbers they are working with increase. For example, they might reason: "There's 2, 4, 6, 8 hands for 4 people, then 9, 10, for 5 people, then 11, 12, for 6 people..." At this point, do not push students to abandon counting by 1's. As their understanding of number grows over the next year or two, they will begin to rely less on counting by 1's. However, they may continue to count by 1's to validate their thinking or to approach challenging problems.

- Can students explain their strategies? How clearly can they record them? Do they use pictures? words? How are they using numbers or equations as part of their recording?

  Ask students to tell you about how they solved the problem as they show you how they have recorded their thinking. Compare what they *say* with what they have *written*. If there is more to their solution than they have recorded, help them figure out ways to write or draw about what they did.

For more challenge, students can find and record more than one way to solve the problem, or they can find the number of hands in a larger group or in the whole class.

Activity

Sharing Strategies

Gather the group together to share strategies for solving the problem. Have counters available for students who want to use them to demonstrate their strategies.

As volunteers share their approaches, record each different method on the board or on chart paper. Use a different color for each method so that students can refer to them easily ("Mine is like the pink way," or "Mine is almost the same as the blue way, but..."). Alternatively, you can letter or number each method. Avoid labeling them with students' names because this takes ownership of the method away from other students who also used it.

Your ways of recording in these whole-group meetings give students models for recording their own work. Whenever possible, follow the student's lead: If the student drew pictures of people, you draw pictures of people; if a student used tallies, use them yourself. Sometimes, you will need to develop your own way to show the student's thinking process. For example, a student who found the solution by modeling the problem with cubes and then counting them might record by drawing a picture of the cubes and the total. To record this strategy for the group, show a picture of the cubes *and* the numbers the student said when counting them (as in example C, below).

Although students will be familiar with addition notation from their work earlier in the unit, they may find it more natural to represent their strategies in other ways. Use equations if students have used them, or if you find them a natural way to record a particular strategy, but treat equations as *just one* of the many good ways of recording that students have used. See the **Teacher Note,** Introducing Notation (p. 162), for more information on the use of equations in this situation and why students need opportunities to find their own ways to record their thinking.

Here are the ways one teacher recorded five different approaches:

Mia: I drew 8 people and counted the hands.

Jacinta: I counted by 2's, for each person.

Chris: I used cubes to show the hands, and I counted them, 2, 4, 6, 8, 9, 10, 11, 12, 13, 14, 15, 16.

Yanni: I wrote down 2 for each person, and I added all the 2's: 2 + 2 is 4, and the next two make 4, and 4 again, and 4. Then 4 + 4 is 8, and 8 here too. And 8 + 8 is 16.

Susanna: I knew 4 people have 8 hands, from before. So, 8 people have 8 + 8 hands. I took 2 from one 8 and put it with the other 8 and got 10. Then there's 6 more from the 8, and 10 and 6 is 16.

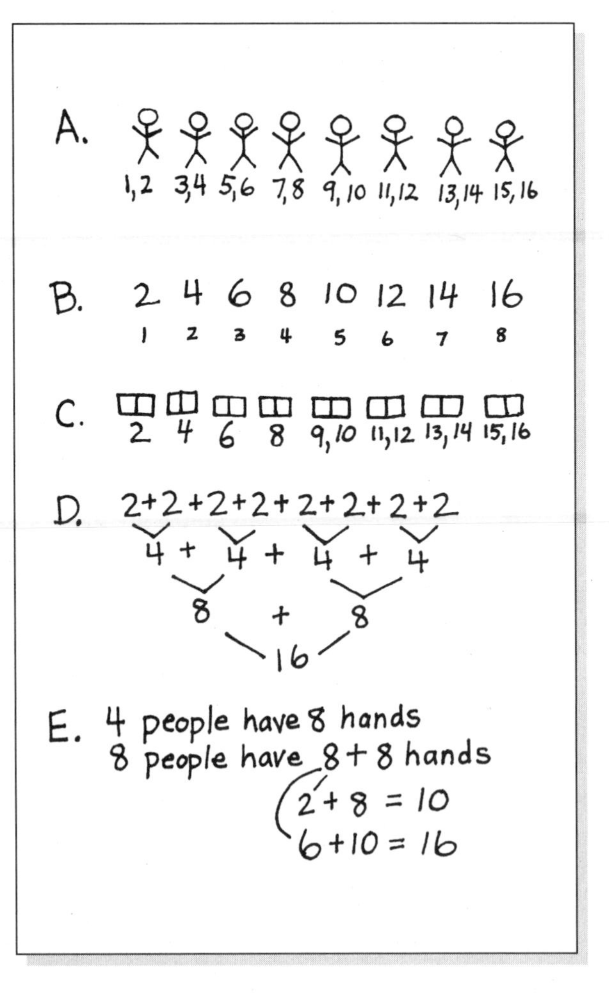

After several students have shared their approaches, ask:

Does anyone have a way that is different from one of the ways I've written here?

When you have recorded five or six different strategies, or when students are becoming less attentive, ask each student to look at his or her own approach and decide which of the ones you've recorded is closest to their own. Ask for a show of hands for each approach. This validates all students' approaches and also gives you a sense of which strategies the class is using.

Activity

Counting Hands in the Group

Suggest counting around a group by 2's as a way of double-checking the hands in the problem they just solved. Call up eight students (or the number you used in the problem). Ask these students to say, in turn, two numbers in sequence, one for each hand. That is, the first says "1, 2," the second "3, 4," and so on. Encourage them to say the first number softly and the second number louder so everyone can hear the counting-by-two sequence while also seeing how it relates to counting a group of two objects.

Counting Around Again Repeat the activity once or twice. Call up a different group each time, until everyone has had a turn.

If you think students are ready, try counting around to find the number of hands in the whole class. It is easy to keep track of who has counted if everyone stands up before the count begins, then each sits down right after counting.

If you count around the whole class more than once, consider counting eyes, ears, or feet on successive counts. Each time, begin with a different student and proceed around the class in a different order. This gives students a chance to hear the counting sequence again and to have their turns at different points in the sequence. Repeating the activity also gives students a chance to think about whether the total will always be the same, no matter how many times you count.

We're going to start counting with Claire this time. What number do you think we'll end on this time? Why do you think so?

The opportunity to count hands, feet, eyes, and so forth, either around the class or a smaller group, comes up again in the next session. This activity can be repeated many times, increasing the group size as students become ready.

Session 1 Follow-Up

Homework

How Many Hands at Home? Send home Student Sheet 10, How Many Hands at Home? Students draw a picture to show all the people who live at home with them, and they find the total number of hands at home. They show how they found their solution, using pictures, numbers, and words, so that someone looking at their work would know exactly what they did.

Students may ask if they can include pets who live with them. Explain that first, they should find only the total number of *hands*. Then, if they like, they may solve a second problem, finding the total number of hands *and paws* at home.

Extensions

How Many Fingers in the Group? Students find the number of fingers in a group of four or five students. They record how they found their solutions.

How Many Feet in the Bed? Read aloud the book *How Many Feet in the Bed?* by Diane Johnson Hamm (Simon and Schuster, 1991). The book shows several different combinations of people and pets in a bed. Pause occasionally while reading and ask students how to find the total number of feet (including paws) in the bed. Students might make up variations of the story, drawing a combination of people and animals in a bed and recording how they know how many feet and paws are in the bed.

Counting Around the Class by 2's The first student to count says 2, the next says 4, the next 6, and so on. Students may use counters or a 100 chart for reference. Counting by 2's can be challenging for many first graders. Be sensitive to competition and potential embarrassment for students who have difficulty figuring out their number. If you think particular students may have difficulty with the later numbers in the count, call on them first, so that they supply the earlier numbers. For extra challenge, students count by 2's, *starting with 1*. That is, the first student to count says 1, the next says 3, the next 5, and so on.

Dealing with Sensitive Issues

Teacher Note

Two sensitive issues may arise in your classroom when students are solving problems about body parts that come in pairs. First, there may be some general silliness as students brainstorm lists of things that come in twos. You will have to handle these remarks according to your own and your school's approach. In field-test classes, students were so engaged in the mathematics of this activity that the discussion of more private parts of the body never arose. However, to be prepared, you might want to consider in advance how to keep the focus on the mathematics.

A second issue may arise in classrooms with a student who does not have the usual number of body parts (for example, a child who is missing an arm or several fingers). If this is the case in your classroom, you will have already talked with your class about the student's special needs. Presumably, some matter-of-factness toward this child's differences will already have been established. In the activity Hands and Other Pairs (p. 49), the student's differences should be included in the discussion as a mathematical issue.

If we are going to count hands and Patrick has one hand, how can we make sure we get the right count?

Depending on your own style and the atmosphere in the class, you can even make this a highlight of the mathematics:

One group of five students had ten hands, but Molly thinks something different might happen for this new group of five students. Who can predict how many hands we will have?

Such discussions are often much more uncomfortable for adults than for students—including the student who is different, who is likely to be genuinely interested in such a problem.

D I A L O G U E   B O X

Four Students, Eight Hands

The teacher calls four students to the front of the room and asks the class to figure out how many hands in this group of four. Students have a variety of strategies, including counting in different ways, doubling, using number combinations, and breaking the problem into smaller parts and combining the parts.

Here are four people. Somebody just said that each person here has two hands. How many hands are there?

Fernando: There's 4. No, there's 4 *pairs.*

Iris: There's 8.

How did you figure that out?

Iris: Because if they just had 1 hand it would equal 4, but they have 2 hands, and 4 and 4 is 8, so they have 8.

So, Iris doubled the number of people. Who did it a different way?

Chris: I took Kaneisha and Diego off for a second, and then Michelle has 2 hands, and Nadia has 2 hands, and I know 2 + 2 equals 4. And then I added on the other people. Kaneisha has 2 hands and Diego has 2 hands. And 2 + 2 is 4, again. Then, 4 plus 4 is 8.

Chris did it in a different way. He broke the problem into two groups of two people. Then he combined his answers.

Brady: I counted by 2's: 2, 4, 6, 8.

Chanthou: I want to add something to that. Even if you count by 2's or if you count by 1's you get the same number.

Can you say more about that?

Chanthou: Brady said 2, 4, 6, 8. Or you can go 1, 2, 3, 4, 5, 6, 7, 8. It's the same number.

That's a good way to check. If you count by 2's, you can count again by 1's, to see if it comes out to the same number. OK, who has a different way?

Tuan: I counted and I went... 1, 2, 3...1, 2, 3... 1, 2.

Can you tell us more about what you counted? What did the first "1, 2, 3" stand for?

Tuan: Well, 1, 2, 3 *[holds up the first three fingers on his left hand],* that's Michelle's two hands and one of Nadia's. Then, 1, 2, 3 *[holds up the remaining two fingers of his left hand and the thumb of his right hand]*—that's Nadia's other hand and Kaneisha's hands. Then, 1, 2 for Diego *[holds up the next two fingers on his right hand].* That's 8.

Libby: You could do it like that except you could make it all into 3's: 3, 3, 3. Three and 3 is 6 and 3 is 9. But you don't have the last hand, so take away the extra one and it's 8.

That's a different way of thinking about it. With hands, many people think in 2's. Libby and Tuan thought about it in 3's. It's great that you all thought about this problem in different ways!

Helping Students Record Their Strategies

As students explore how many hands there are in a group of eight students, the teacher helps them through some typical difficulties as they try to record their solution strategies.

When the teacher arrives at Andre's desk, she sees that he has recorded only the number 16.

Can you tell me what you did to get 16?

Andre: I did it in my head.

Can you tell me how you did it in your head?

Andre: I knew it was 16. It was easy.

How did 16 just pop into your head? What were you thinking about?

Andre: I don't know… 8 and 8 is 16.

Eight? What did the 8 come from?

Andre: Four kids have 8 hands. From what we did before.

That's great—you used what you learned in one problem to help find an answer to another one. So, you could write down that you learned that 4 people have 8 hands.

Andre: Just write it down?

Yes, if you just learned it, you can say so. We're always using things we learn to help us solve new problems. So, tell me again what you did to get 16. You knew 4 people have 8 hands. Then what?

Andre: I needed how many hands for 8 people, and 8 and 8 is 16.

How did you figure out 8 and 8 is 16?

Andre: I didn't. I knew that, too.

OK, so, you can find a way to write that down, too.

With some prompting from the teacher, Andre was able to explain his solution strategy. The teacher also helped him to understand what is appropriate to record: it's OK to write down that you "just know" a number combination, but you need to explain how you used that combination to help you arrive at the solution.

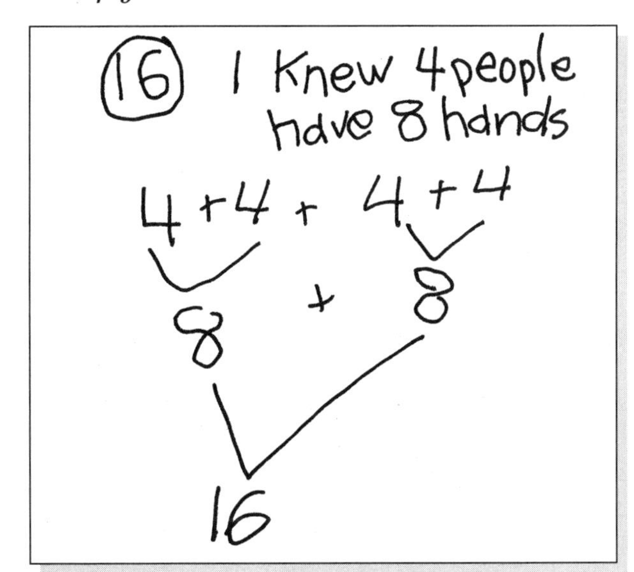

Andre's work

The teacher moves on to a student who has recorded only part of her solution strategy. Kaneisha made eight pairs of cubes, and then counted each cube. She drew her arrangement of cubes, wrote 16, and said she was finished.

Can you tell me about what you drew?

Kaneisha: It shows I used cubes.

What did you do with the cubes?

Kaneisha: I put 2's of them together.

How did putting the cubes in pairs help you?

Kaneisha: I counted them, like, this one *[points to a pair of cubes]* is someone. It's someone's hands, and there's 8 people.

Can you show me how you counted them?

[Kaneisha touches each square as she counts.]

So, your picture shows that you used cubes for the 8 people, and that you got 16 hands. What if I'm reading this later. How could I know exactly how you counted those squares, what

Continued on next page

numbers you said, and what squares they stood for? You started out saying "one." What did you count when you said "one"?

Kaneisha: This one *[points to a square].*

So, how could you record that?

Kaneisha: I could write 1 here *[in the square]*?

If that's what you said when you counted it.

Kaneisha: Then 2?

The teacher nods, and Kaneisha continues counting each square and recording the number she says. Although Kaneisha could readily tell the teacher how she solved the problem, like Andre, she needed some help understanding what is appropriate to record.

Kaneisha's work

Sometimes, a student's recording method doesn't accurately reflect the solution strategy. For example, earlier in the session the teacher observed Chanthou explaining her approach to a classmate.

Chanthou: You can go *[keeps track on her fingers as she counts]* 2, 4, 6, 8, 10, 12, 14, 16. That's 8 numbers *[holds up 8 fingers]* so there are 16 hands.

At that point, Chanthou had not yet recorded her work. When the teacher returns a little later, Chanthou has drawn a picture of 8 stick figures, each with 2 hands, and numbered the hands 1 to 16.

Can you tell me what you did to solve the problem?

Chanthou: I drew a picture of 8 people and counted the hands.

That's a great way to solve this problem—you can draw it out and count it up. Is there another way you could solve this problem? Could you count in a different way?

Chanthou: I'm not sure.

Well, how many hands does one person have?

Chanthou: Two. Oh... You could go... 2, 4, 6, 8, 10, 12, 14, 16.

So, you have two ways to solve this problem. See if you can find a way to show how you counted by 2's.

Chanthou's work

Occasionally, when students have developed a recording method they are comfortable with, they continue to use it even if it does not reflect their thinking. They may not be fully aware of what strategies they used, or they may not completely trust the new strategies and may feel more comfortable recording a more familiar approach. You can help these students become more aware of their thinking and more confident by encouraging them to find and record more than one solution strategy.

Twos and Fours

What Happens

The class lists and makes predictions about the number of hands for successive numbers of people. Then students find the total of several 4's in a problem about wheels on cars. They record and share their solution strategies. Student work focuses on:

- exploring patterns in even numbers
- developing meaning for counting by 2's
- finding the total of several 2's or 4's
- recording strategies for counting and combining, using pictures, numbers, and words

Materials

- Unlined paper
- Student Sheet 11 (1 per student, homework)

Activity

How Many Hands at Home?

At the start of the session, take about 15 minutes for sharing the completed homework, How Many Hands at Home? Call on a student to give the number of people at home, and ask the others in the class to find the number of hands.

Susanna has 6 people in her house. Does anyone else have 6 people at home? How many hands do 6 people have? How do you know? Who has another way to figure it out?

Gather several different strategies for finding the number of hands. Repeat the activity once or twice, asking for a different number of people at home each time.

Activity

Patterns in People and Hands

Set up a two-column chart on the board or chart paper, labeling the first column *People* and the second *Hands.* Use this chart as students make predictions about the number of hands for successive numbers of people.

Let's think about the number of hands for different numbers of people. What if there was just one person at home? How many hands?... OK, so 1 person, 2 hands. How many hands for two people at home? How do you know?

Continue to gather and record numbers of hands for up to about 6 people. Ask students to explain how they arrived at their answers. Some students may begin to recognize or extend patterns. ("We had 10 hands for 5 people, so for 6 people, it goes up 2 more, so that's 12.") Others will treat each new number of people as a separate problem, unrelated to the previous ones.

After about 6 people, jump to a slightly larger number:

OK, so 6 people have 12 hands. What about 9 people? How many hands would they have? How do you know?

Give students a couple of minutes to find the solution. They may use counters or pencil and paper, but they do not need to record their strategies. Then, ask a few to share their solution strategies. When you record the number on the chart for 9 people, leave space to record any numbers you have skipped, since students may give these as they share their strategies.

If you think that many students are noticing patterns, ask about other larger numbers of people, such as 11 or 12.

Take a few minutes for students to count the hands (or eyes, or ears, or feet) in the whole class. If you have a very large class, you may want to specify a smaller group for students to count around. Some students may find it helpful to refer to the class chart of people and hands. Repeat the counting once or twice, each time with a different body part that comes in pairs.

Activity

A Problem About Fours

We've been talking about hands and other things that come in pairs. Now, we're going to solve a problem about something that comes in *fours*. What do you know that comes in fours?

Take a few minutes to brainstorm with students. Suggestions commonly include paws on animals, wheels on various vehicles, and legs on tables and chairs.

❖ **Tip for the Linguistically Diverse Classroom** Record students' ideas on the board with quick sketches beside the words. For the problem that follows, have available a toy car with wheels as a visual reference.

Write this problem on the board:

Mia has 3 toy cars. How many wheels are there?

Read the problem with the class. Explain that students are to solve this problem on paper, recording how they found their solutions. Students may work alone or in pairs, but each student should record his or her own solution strategy.

Note: The numbers in the problem are intentionally small to encourage a wide range of approaches. When solving problems with small numbers, some students may be able to use familiar number combinations and relationships in finding the solution. When solving similar problems with larger numbers, these same students may return to counting by 1's. If some students solve this problem easily and record their work well, ask them to find the number of wheels on 6 or 7 cars.

Observing the Students

- What strategies do students use? Do they count out each quantity in the problem? Do they count by 2's or 4's? Do they use strategies that involve number combinations? doubles? breaking the problem into parts and then combining the parts?

  Encourage any students having difficulty to model the problem with counters or pictures.

- How do students record their work? Are they using pictures? words? How are they using numbers or equations as part of their recording?

Keep in mind that recording problem-solving strategies is challenging for young students. You may need to continue to work with some students throughout the unit to help them find ways to explain and record their thinking. Students who finish early can find and record more than one way to solve the problem.

When everyone has finished, ask for volunteers to explain how they solved the problem. Gather a few different solution strategies. Have counters available for students who want to use them to demonstrate their strategies. As you did when students shared strategies in Session 1, record each method on the chalkboard or on chart paper. Whenever possible, base your model on the way the student recorded.

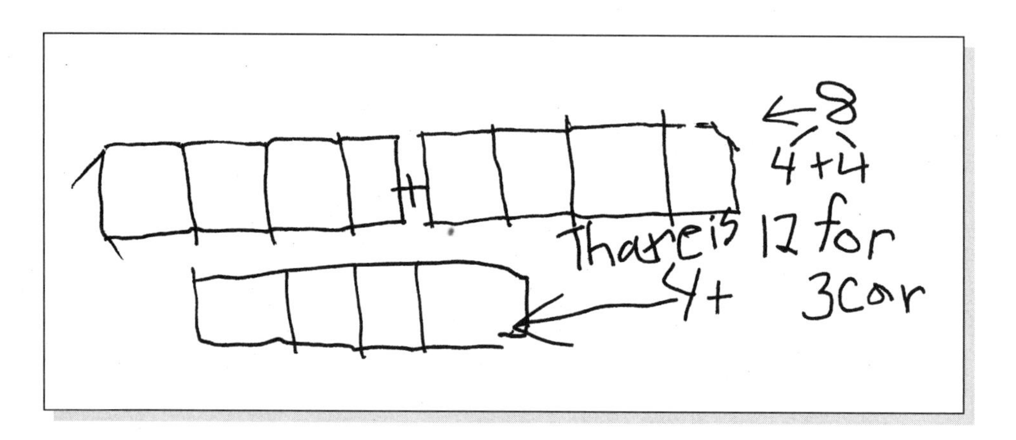

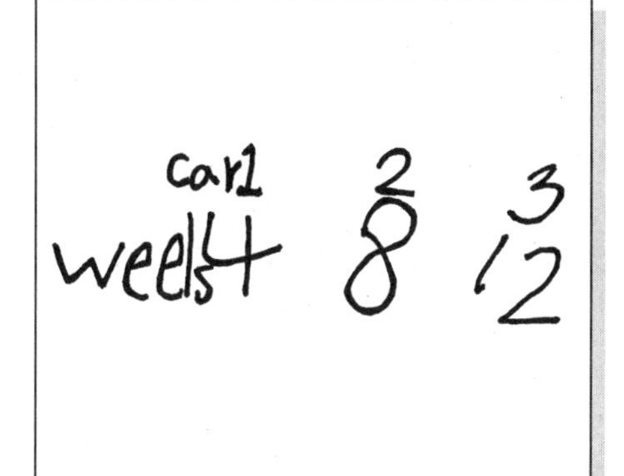

Session 2 Follow-Up

Cats and Paws Students solve the story problem on Student Sheet 11, Cats and Paws, and they record their solution strategies.

❖ **Tip for the Linguistically Diverse Classroom** Read the problem aloud. Students add their own rebus drawings to the sheet to help them remember key words *(cats, paws).*

What Comes in Twos? Students make their own illustrated lists of things that come in groups of different sizes. They might then read the book *What Comes in 2's, 3's, and 4's?* by Suzanne Aker (Simon and Schuster, 1990) for more ideas.

What Else Has As Many Hands? Students find combinations of people and animals that have the same number of hands as their family. For example, 4 people have 8 feet, and so do 2 people and 1 dog.

Session 3

Collect 25¢ Together

Materials

- Penny, nickel, dime, and quarter (1 each per student, in a small container)
- Prepared coin sets (1 per pair)
- Dot cubes (1 per pair)
- Student Sheet 12 (1 per student, homework)

What Happens

Students briefly explore coins, then play the game Collect 25¢ Together, rolling a dot cube and collecting coins until they have a total of 25¢. Their work focuses on:

- counting and keeping track of about 25 objects
- becoming familiar with coins and equivalencies among them

Activity

Exploring Coins

First graders vary widely in their experience with and knowledge about coins. Some have little understanding of coins; some know coin names and values but not coin equivalencies; and some are comfortable making trades among coins of different values.

Pass out a penny, nickel, and dime to each student. If possible, distribute these in a small container, such as a paper cup.

Ask students what they notice about the penny. You might give them a moment to talk with a neighbor before they share their observations.

Everyone hold up a penny. Look around you. Do you see all pennies? Look closely at your penny. What do you notice about it? Tuan says it's a red-brown color. Who notices something else?

You might decide to record students' ideas on chart paper. Some students will probably talk about coin values and relationships among coins, while others might focus on coin appearance or the things coins can be used for. This variation is illustrated in the **Dialogue Box,** What Do You Notice About Coins? (p. 69). All observations help students become more familiar with coins.

Repeat the process with a nickel, then again with a dime, and finally with a quarter. Then ask students about similarities and differences among the coins.

What things are the same about all the coins? How is the penny different from the others? What's different about the nickel? What about the dime? the quarter?

If students get restless, you can introduce the quarter at another time, since they won't need it in the following game.

Activity

Collect 25¢ Together

Collect 25¢ Together is a cooperative game that involves collecting at least 25¢ together. Introduce the game by asking two students to play a few demonstration rounds. They will need one dot cube and one of the prepared coin sets. Start by playing with just pennies, no trades; that way, students who are not yet comfortable with coin equivalencies can still follow the game. (See the **Teacher Note,** Coin Equivalencies in Grade 1, p. 68, for more on this topic.) Later, when students play in pairs, they may either work with pennies only, or they may trade coins for equivalent amounts at any time.

Today you will play a game called Collect 25¢ Together. In this game, you and a partner work together to collect 25 cents.

Briefly explain that players take turns. On each turn a player rolls a dot cube, collects the number rolled in coins, and determines the total amount collected so far. Throughout the demonstration game, involve the whole class in thinking about how much money has been collected so far and how they know.

Eva just rolled 5. How much money does Eva take? Now it's Tony's turn. He rolled 3. How much does he take? How much do they have in all now? How do you know?

Play for several turns or until you think students understand the game. If some students raise the possibility of using coins other than pennies, encourage them to explain how they would do this.

Why did you say you could take a nickel and a penny for a roll of 6? Is that the same as 6 pennies?

If no one suggests making trades during the demonstration game, raise this possibility. Count out five pennies the players collected and ask if anyone knows a coin worth 5 cents that they could trade the pennies for.

Explain that when playing in pairs, students may trade pennies for other coins if they want, or they may play the game with just pennies.

For the rest of Session 3, students play in pairs. If necessary, three students can play together. Each pair needs a dot cube and one of the prepared coin sets. Point out that they need not get exactly 25 cents. If their last roll gives them more than 25 cents, that's a perfectly good way to finish the game.

Observing the Students

As students play, circulate to observe their strategies and to offer support as needed.

- How comfortable and accurate are students as they count the coins? What sorts of errors do you notice in their counting? How do they keep track of what has been counted and what needs to be counted?

  If some students are having difficulty, ask them to collect a smaller amount, such as 15¢.
- How do they find the total after the new coins have been added? Do they count all the coins? count on from the number they had for the previous turn? use their knowledge of addition combinations? count by 5's or 10's? Can they determine how much more they need for a total of 25 cents?
- What coins are they familiar with? Do they know coin names? coin values? any equivalencies?

- Do they collect rolls of 5 or 6 in pennies, or do they use nickels? Do they trade any of the pennies they have collected for coins of equal value? Do they count these coins accurately?
- Do they recognize when they have 25 cents? Can they determine how many more than 25 cents they have at the end of a game?
- Do students play cooperatively and help one another with the game? Sometimes a student who is not yet comfortable making trades will work well with one who is, as long as both students explain their thinking and listen to one another. At other times, it may be better to make pairs of students who are comfortable making trades and pairs of students who are using just pennies. As needed, make adjustments in the pairs as you observe them playing.

If some students are ready for more challenge, they can collect a larger amount, such as 50 cents. If these students are not yet comfortable making trades, they will need more pennies added to their set.

Session 3 Follow-Up

Collect 25¢ Together Send home Student Sheet 12, Collect 25¢ Together. To play the game with someone at home, students will need a dot cube and a collection of real or play coins. Families may have a dot cube at home (from another game). If not, they can use Number Cards 1–6 or write the numbers 1 to 6 on slips of paper and draw them randomly from a pool.

Ways to Make 15¢ Students find different coin combinations that make up 15 cents. For example, they could use 15 pennies, a dime and 5 pennies, 3 nickels, and so forth. They record their solutions. For more challenge, they can find different combinations to make up 25 cents.

Teacher Note — *Coin Equivalencies in Grade 1*

In this investigation, students work with coins as they play the game Collect 25¢ Together. Your students will probably vary widely in their understanding of coin equivalencies. While some students may readily trade coins, many will not initially be ready to make trades and will work with just pennies. As they play repeatedly, they will learn some other coin names and values and begin to make a few simple trades, such as 5 pennies for a nickel. But many will continue to rely on pennies most of the time.

Some students may be inconsistent in the ways they count and make trades. Counting up to 25 objects can be challenging for first graders. They may find it hard to keep in mind coin values and relationships while also concentrating on counting and keeping track of their growing collection. For example, some students may accurately count a small set of coins, such as a nickel and two pennies ("5 and 2, that's 7 cents"), but with a larger set, such as 5 nickels and 15 pennies, may mistakenly count some of the nickels as "ones." Some may comfortably trade a set of 5 pennies for a nickel, but have difficulty trading a set of 20 pennies for nickels. And some students may tell you that they know a nickel is worth 5 pennies and a dime is worth 10 pennies, but nonetheless count each coin in a set of pennies, nickels, and dimes as "one." All these students will need more experiences, both with coins and with counting objects, before they are able to count a large set of coins accurately.

As students work with coins in this investigation, do not insist they make trades, but encourage them to explore possible trades if you think they are ready.

Is there something you could use instead of 5 pennies? How else could you make 7 cents?

Students will have many opportunities to continue developing their understanding of coins and coin equivalencies the second grade *Investigations* curriculum. Over time, as they work with coins in many different ways, they will gain a deeper understanding of coin equivalencies and will become comfortable making up amounts of money in different ways.

What Do You Notice About Coins?

These students are looking at plastic pennies, nickels, dimes, and quarters. For each coin, the teacher begins with an open-ended question and then follows students' lead in choosing follow-up questions. Students' observations range widely, from how coins look and what they are used for, to more specific comments on coin relationships and calculations involving coins.

Is this real money? *[Students answer "No!" in chorus.]* **You'll notice a lot of things that are the same in real and plastic money. Everyone hold up a penny. What do you know about pennies?**

Donte: They're one cent.

Libby: They're brown.

Nadia: If you put 10 pennies together, that'll make 10 cents.

Tony: A picture of Abraham Lincoln is on the front.

Eva: Sometimes they are shiny and sometimes they are dirty.

Which ones are shiny?

Diego: The new ones.

Which ones are dirty?

Claire: Older ones.

What about nickels? Everyone hold up a nickel.

Jacinta: Five cents.

Nadia: If you put 2 nickels together, it'll make 10 cents. Five cents plus 5 cents equals 10 cents.

Kaneisha: You can't buy anything with a nickel.

[Several students object.]

What can you buy with a nickel?

Andre: Maybe a little piece of candy or gum.

Iris: If you put three nickels together you'll have *[pause]*... 16 cents.

What's a nickel worth?

Iris: Ten. No, 5.

So, three 5's.

Iris: Five and 5 is 10. Then *[counts on her fingers]* 11, 12, 13, 14, 15!

Eva: It's 15 because you can count by 5's: 5, 10, 15.

You also have dimes. Everyone hold up a dime. What can you say about dimes?

Yukiko: That's 10 cents.

Luis: There's a man on the front.

Shavonne: It's the smallest money.

It is the smallest coin.

Shavonne: How come this one [a dime] is small and it's 10 and that one [a nickel] is big and it's 5?

Nadia: But 25 cents *[holding up a quarter]* is bigger than the others and it's more.

[The teacher holds up a nickel and a dime.] **Why do you suppose one is bigger than the other?**

Shavonne: If you thought 5 cents is 10 cents... if they were the same size, you wouldn't know which is which.

So the different sizes help us tell them apart.

Mia: There's a tree in the back [of the dime].

Garrett: And candles.

Kristi Ann: If you have 10 of these, you'll have 100 cents.

Imagine if you had 100 pennies every time you had 100 cents. That would be very heavy. What do people carry that's easier to carry?

William: You could put them in bags.

What else is worth 100 pennies?

Nadia: Four quarters.

Yanni: One dollar.

It's a lot easier to carry a dollar bill than 100 pennies.

Sessions 4 and 5

Counting and Combining

Materials

- Student Sheet 13 (1 per student)
- Prepared coin sets (1 per pair)
- Dot cubes (1 per pair)
- Transparencies of Quick Image Squares
- Squares, Sets A and B (4 of each, in envelopes)
- Transparencies of loose squares (singles and pairs)
- Interlocking cubes (15–20 per student)
- Overhead projector

What Happens

Students do Quick Images with pictures of squares arranged in groups. In How Many Squares? they find the total number of squares in two sets of squares grouped by ones, twos, and fives. During Choice Time, they continue work on How Many Squares?, Collect 25¢ Together, and solving problems with 2's, 4's, or 5's. At the end of Choice Time, they share strategies for solving one of these problems. Students' work focuses on:

- developing strategies for organizing sets of objects so that they are easy to count and combine
- recording strategies for counting and combining, using pictures, numbers, and words
- finding the total of several 2's, 4's, or 5's
- becoming familiar with coins and equivalencies among them

Activity

Quick Images with Squares

Gather students where they can see images projected from the overhead. Distribute interlocking cubes to tables or groups so that each student has access to 15–20 cubes. Explain that you will be showing some Quick Images, but that this time, the images will show patterns of *squares* instead of dots. Students can either draw the images they see or construct them with interlocking cubes.

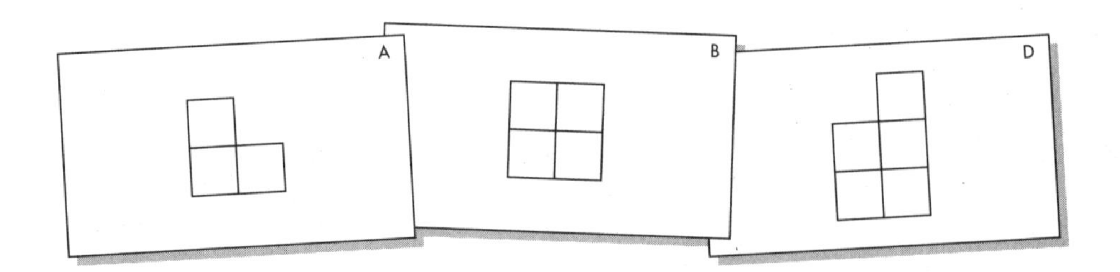

Start with image A; then use two or three others (such as B, D, and F).

See p. 7 to review the steps in Quick Images. After doing Quick Images with dot patterns, students may find the square patterns a bit challenging at first. At least initially, plan to flash each image for a couple of seconds longer than you did the dot patterns.

Activity

Introducing How Many Squares?

Place transparencies of single squares and pairs of squares on the overhead in a random arrangement. Use three or four singles and three or four pairs, for a total of about 10 to 12 squares.

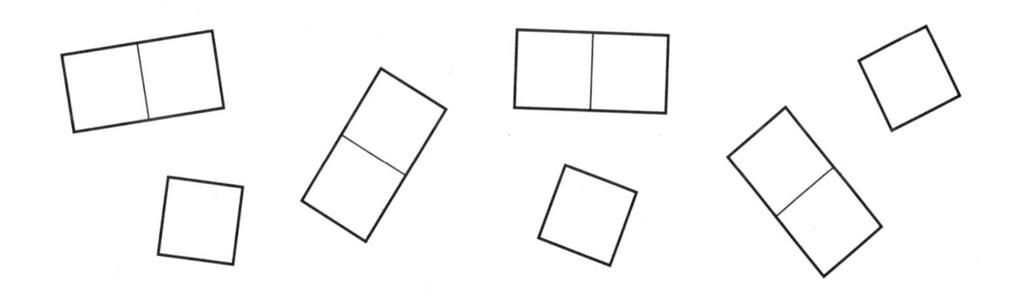

Ask students to find how many squares there are. Then ask for volunteers to demonstrate their counting strategies at the overhead. If a student's method involves organizing the squares in some way, mix them up again before the next student shares a strategy. If a student does not demonstrate organizing the squares to help count and keep track of them, model this yourself if it seems natural to the student's approach. For example, if a student explains that she counted the pairs by 2's, you might move the pairs off to the side, lining them up as she counts them.

If your students cannot think of more than one or two ways to count the squares, help them out by arranging the squares in a way that may suggest a new counting strategy, such as in groups of two or in rows of five. See the **Dialogue Box,** Finding Ways to Count (p. 76), for the arrangements one teacher used to help her class.

Explain that when they do How Many Squares? as a Choice Time activity, students will count sets of squares like this one, provided in envelopes. They will need to make a record on paper of how they found the total number of squares, including a picture of how they arranged the squares, and an explanation in numbers or words of exactly what they did to find the total.

Activity

Choice Time

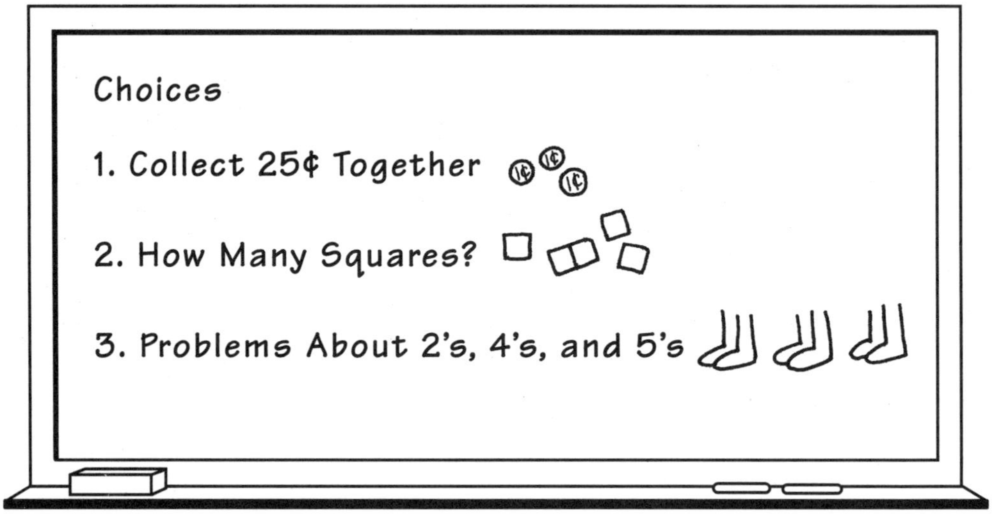

Choices

1. Collect 25¢ Together

2. How Many Squares?

3. Problems About 2's, 4's, and 5's

Post a list of the three choices students will work on for the rest of today and most of next session. All students need to have completed their work on the story problems before the last 20 minutes of Session 5, when they will share their strategies for one of the problems on the sheet.

Choice 1: Collect 25¢ Together

Materials: Dot cubes (1 per pair); prepared coin sets (1 per pair)

In this game, pairs (or if necessary, three students) work cooperatively to collect 25 cents. Each pair needs one dot cube and a set of coins. The first player rolls the dot cube, reads the number rolled as *cents,* collects that amount in coins, and sets out these coins to begin the pair's coin collection. The other player then repeats the procedure, adding more coins to their mutual collection. Students may choose to take only pennies, or they may trade coins for equivalent amounts at any time. Players continue taking turns until they have 25 cents or just over (they do not need to collect 25 cents exactly).

If you noticed that some students had difficulty working with 25 cents when the game was introduced in Session 3, suggest they play to collect a smaller amount of money, such as 15 cents. For more challenge, students can play to collect a larger amount, such as 50 cents. (If these students are not comfortable trading, they will need more pennies in their coin set.)

If you see many students begining to make trades, you might call everyone together briefly to play a class game, focusing on possible trades. As you did when introducing the game, choose two volunteers to play, but involve the whole class in each turn. Encourage students to share their thinking about possible trades, but also recognize that some students may not follow the trading and will continue to use pennies only when they play.

Choice 2: How Many Squares?

Materials: Squares, Sets A and B (four of each set, in separate envelopes); lined or unlined paper

Students take a set of Squares and work with a partner to count the squares in that set. Set A contains single squares and pairs, and Set B contains single squares and five-square strips. Students may start with either set. On a sheet of paper, students record the letter of the set they are counting (A or B), and they make a record of how they found the total number of squares. This record should include a picture of how they arranged the squares, and an explanation, with numbers or words, of exactly what they did to find the total number of squares.

Remind students to put the set of Squares back in its envelope and return it before taking the other set.

For more challenge, students can find and record more than one way to find the total number of squares. Or, you can provide a larger set of squares and ask them to find the total.

Choice 3: Problems About 2's, 4's, and 5's

Materials: Student Sheet 13, Feet, Fingers, and Legs (1 per student)

Students solve the problems on Student Sheet 13, Feet, Fingers, and Legs, in which they find the total of several equal amounts.

❖ **Tip for the Linguistically Diverse Classroom** Read aloud the four problems. Students might add their own drawings to help them remember key words *(people, children, horses, feet, fingers, legs).*

Students may work alone or in pairs, but each should record his or her own solution strategy. They may use pictures, words, and numbers in any combination.

For more challenge, students can find and record more than one way to solve the problem.

Observing the Students

See pp. 66–67 for guidelines on observing students' work during Choice 1: Collect 25¢ Together.

How Many Squares?

- Do students organize the squares in some way to help them count and keep track, or do they count them in a random order? Do they organize the squares by type (all the singles together, all the pairs together)? Do they organize the squares into equal groups, such as groups of four or five?

  If some students are having difficulty counting and keeping track of the squares, you might give them a smaller set of squares to count, such as two or three pairs and several individual squares.

- How do students find the total number? Do they count by 1's? Do they count by numbers other than 1? Do they count the squares appropriately (single squares as 1, pairs as 2, five-strips as 5)? Do they use strategies involving number combinations, such as putting the squares in groups and summing the size of the groups?

  Even when students organize the squares into equal groups, they may still count the squares by 1's. Organizing the squares into groups helps them keep track as they count, but they may not yet be ready to use a single number to stand for a group of objects. Some students, rather than counting by 2's or 5's (the group size), might add all the groups. ("These two 2's are 4, 4 and 2 more is 6, 6 and 2 is 8... [and so on].") They see each group as a 2, and they may even know some numbers in the counting by 2's sequence, but they do not yet see that they can count up by 2 to refer to an increase of 2 more. For these students, addition is still a more comfortable and reliable way of finding a total. With more experiences counting and organizing collections over the next year or two, students will gradually build an understanding of counting by numbers other than 1.

- How do students record their work? Do their pictures of their arrangements of squares show the correct number of squares? Do their pictures accurately show the way the squares are arranged? How clearly can they show their strategies? How are students using numbers to show how they found the total?

  If some students are having difficulty knowing what to record, ask them to show you how they solved the problem. Help them to think about ways to write or draw about what they did. For example, if a student organized the squares into groups of five and then used number combinations to find the total, encourage him to record his reasoning.

  So, you said you knew each of these was a 5, and 5 and 5 is 10. How could you put that down on paper so someone would know that's how you were thinking about it?

For examples of ways that students recorded their work for a similar activity used later in this investigation, see the **Teacher Note**, Assessment: How Many Squares? (p. 95).

Problems about 2's, 4's, and 5's

- What strategies do students use to solve the problems? Do they use strategies that involve counting by 1's? by 2's? by some other number? Do they use strategies that involve number combinations? doubles? breaking the problem into parts and combining the parts?

  If any students are having difficulty, encourage them to model the problem with counters or pictures. If necessary, adjust the numbers in the problem so that students are working with smaller numbers.

  Make note of the variety of strategies you observe students using. When you choose one of these problems as a focus for class discussion at the end of Session 5, find one for which students have used a wide range of strategies.

- How do students record their work? Are they using pictures? words? How are they using numbers or equations as part of their recording?

  If some students make elaborate representations of the situation, ask them to talk about how they would use their drawings to find a solution to the problem. While making detailed drawings can be an important way to make sense of and think through the problem situation, it can sometimes be a way to delay the work of solving the problem.

Activity

Sharing Solution Strategies

About 20 minutes before the end of Session 5, call students together to share strategies for solving one of the problems on Student Sheet 13, Feet, Fingers, and Legs. Have counters available for students who want to use them for demonstration. As you gather a few different solution strategies, record each method on the chalkboard or chart paper, basing your recording on the student's own approach where possible.

Sessions 4 and 5 Follow-Up

Squares in Threes and Sixes To extend the How Many Squares? activity, provide sets of squares that include singles and strips of three, or singles and strips of six. For more challenge, include singles, threes, and sixes in the same set.

DIALOGUE BOX

Finding Ways to Count

To introduce How Many Squares? (p. 71), the teacher arranges a set of squares, singles and pairs, on the overhead and asks students to suggest strategies for counting them. When students do not readily think of different ways, the teacher models a few ways of organizing the squares. As students count the organized squares, they begin to find their own ways of organizing to make them easier to count.

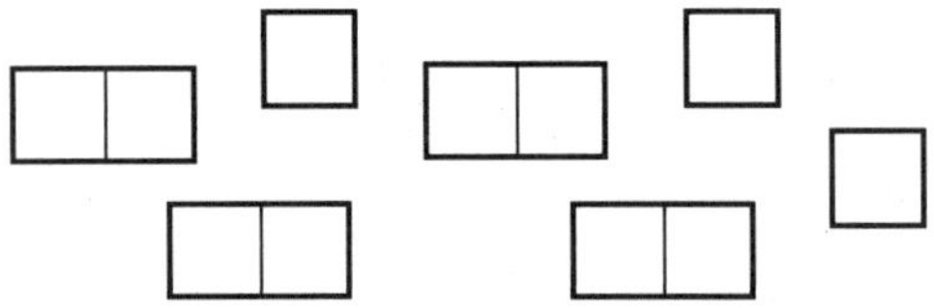

Look at these squares. I'd like you to figure out how many there are altogether.

Jacinta: Ten.

Come up and show us how you counted them.

Jacinta: *Well, 2 [points to the top left pair], 4 [points the bottom left pair]...* 6 *[the top right pair]...* Uh, did I do this one yet *[the bottom right pair]*?

Why don't you try pushing the squares over to the side after you count them? Start again.

Jacinta counts by 2's as she points to each pair in turn and then pushes it into a column at the left side. She then pushes two single squares together, says "10," and slides them together below the other pairs. Finally, she points to the remaining single square and says "11."

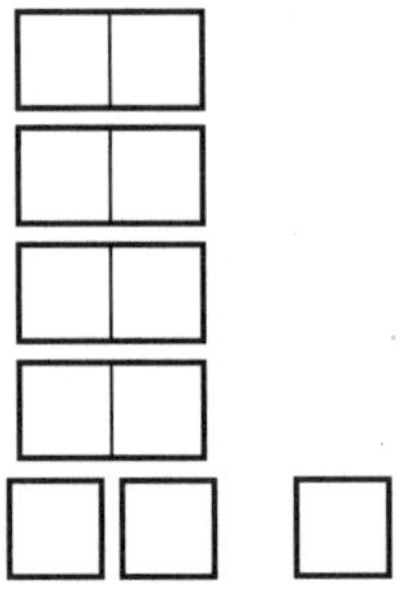

Jacinta, tell us how you counted the squares.

Jacinta: I counted by 2's, and then 1 for this one *[the single square alone].*

That's a great way. Let's see if we can find another way. *[The teacher rearranges the squares as they were initially.]* **Jacinta counted by 2's. Does anyone have another way to count the squares?**

Libby: You can just count them, each one.

Libby demonstrates, pointing to but not moving each square as she counts it. She loses track partway through the count and starts over, this time counting more slowly, to 11.

So, Libby found another way to count the squares. She counted each one. Does anyone have another way? *[After a moment, with no response, the teacher models another way.]* **I'm going organize them in a different way, a little like what Jacinta did.**

The teacher regroups the squares as follows:

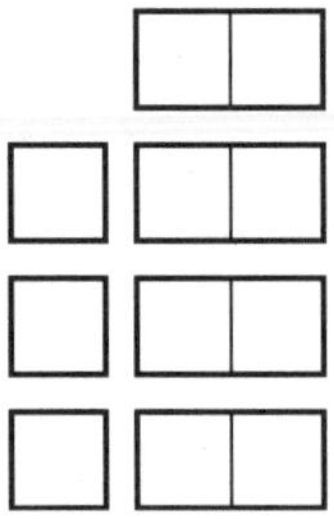

Is anyone thinking about them in a different way now?

After a moment, a few students raise their hands. Jonah comes up to the overhead to demonstrate his way.

Jonah: There's 3 *[points to the singles],* 5 *[the top pair],* 7 *[going down the column],* 9, 11.

So Jonah counted *[pointing to each group]* **3, 5, 7, 9, 11. Does anyone have another way?**

Leah: You could do 3's. *[She comes up to demonstrate, pointing to each row in turn from the bottom up]* 3, 6, 9, and 2 more is 11.

That's another way, looking at the rows: 3, 6, 9, and 2 more is 11. Any others?... OK, let's try one more way to organize the squares.

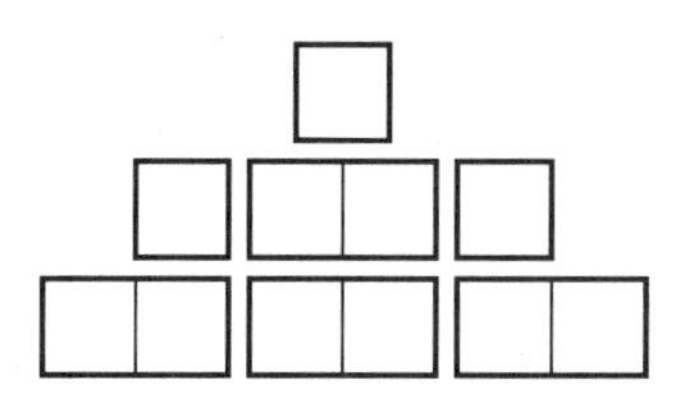

[This time, several students raise their hands right away.]

Donte: You can just do the whole row like Leah's way. The bottom one is 6, then there's 4, and 1.

So 6, 4, and 1. How do you find out how many in all?

Donte: You can go 6 plus 4 is 10, 10 plus 1 is 11.

Diego: I did it sort of like that, but different. I did 6 on the bottom, but 5 on the top, because it's 2 together and 3. So 5 and 6 is 11.

Michelle: When Diego said 5 and 5, I thought of making 5's. Can I show? *[She comes to the overhead and switches a single square in the middle row with a pair on the bottom row, so that the bottom two rows each have 5 squares.]*

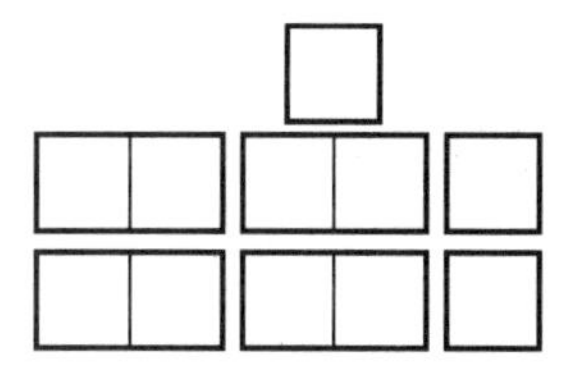

Michelle: So now you can count 5, 10, and 1 more is 11.

Michelle had an idea for a new way to arrange the squares to make them easier to count. Does anyone else have an idea on how we could arrange them?

Many students are now eager to share their ideas. The teacher's modeling a couple of ways seems to have helped students generate their own ideas about how the squares can be organized and counted.

Sessions 6, 7, and 8

Numbers to 100

Materials

- Hundred Number Wall Chart with red chart markers
- Plastic Hundred Number Boards with removable tiles (2 or 3 for the class)
- Prepared coin sets (1 per pair)
- Dot cubes (1 per pair)
- Squares, Sets A and B (4 of each, in envelopes)
- Transparencies of Quick Image Squares
- Overhead projector
- Student Sheet 14 (1 per student, homework)

What Happens

The class plays Missing Numbers, identifying numbers that are missing from a 100 chart. This activity is then added to Choice Time, which continues over the three sessions. At the end of Choice Time, the class repeats Quick Images with squares and shares strategies for How Many Squares? Their work focuses on:

- reading and sequencing numbers to 100
- exploring patterns in the number sequence
- developing strategies for organizing sets of objects so that they are easy to count and combine
- finding the total of several 2's or 5's
- becoming familiar with coins and equivalencies among them
- recording strategies for counting and combining, using pictures, numbers, and words

Activity

Exploring the 100 Chart

If your students are already familiar with the 100 chart, give them time to talk about what they remember about it, to point out any patterns they notice, and to share any new observations. If your students are unfamiliar with the 100 chart, allow a bit more time for exploring it.

Gather students where everyone can easily see the Hundred Number Wall Chart.

Who remembers working with the 100 chart? What do you remember about it?... Yes, the numbers down this right column start with 1, 2, 3, 4. Do you see that pattern down any other columns? What other patterns do you see?

After a few minutes, ask students to locate some numbers on the chart.

Who can find 40? Can someone point to 90? Can someone point to 65? Where is 47?

As students locate numbers, observe how familiar they are with the sequence of written numbers. Do they have a general idea of where to find a number like 47? Do they search the chart randomly? Do they count by 1's until they reach it? If many students are not comfortable with written numbers larger than 50, focus most of your questions on numbers up to 50.

Note: Many teachers use number lines to explore numbers and number patterns in first grade. The *Investigations* curriculum uses the 100 chart because the arrangement of numbers in rows of 10 helps students attend to the patterns of tens and ones in our number system. Both the number line and the 100 chart are useful tools, and you may choose to make both available for student reference when they need to combine, separate, and compare quantities. However, the activities Missing Numbers (following) and Clapping Patterns (Session 9) specifically require use of a 100 chart; a number line cannot be substituted.

1	2	3	4	5	6	7	8	9	10
11	12	13	14	15	16	17	18	19	20
21	22	23	24	25	26	27	28	29	30
31	32	33	34	35	36	37	38	39	40
41	42	43	44	45	46	47	48	49	50
51	52	53	54	55	56	57	58	59	60
61	62	63	64	65	66	67	68	69	70
71	72	73	74	75	76	77	78	79	80
81	82	83	84	85	86	87	88	89	90
91	92	93	94	95	96	97	98	99	100

Activity

Missing Numbers

Missing Numbers, a game that involves locating numbers on the 100 chart, requires only a brief introduction. (Students may remember it from the unit *Building Number Sense.*) After playing the game as a whole-class activity, pairs will play together during Choice Time in Sessions 6–8.

Remove the cards 22, 24, 26, 28, and 30 from the Hundred Number Wall Chart and put them facedown out of sight. Have the red chart markers at hand and use them to cover the numbers you replace in the slots as students identify them. Point to any one of the five empty slots.

Think silently, in your head, about what number is missing from this spot. When you think you know what it is, don't say it or raise your hand—just look at me like you're ready to tell me.

Pause long enough to give everyone a chance to think about what number is missing. Then ask for several students to explain what number they think is missing and why.

Why do you think it's 24? Chris says he thinks it's 24 because it comes after 23. Who has another reason?... Fernando counted up from 21, at the start of the row. Let's try counting that way together.

Who found the missing number in another way?

When no one can suggest another way, ask a volunteer to write the missing number on the board. Then, put the 24 in the slot and cover it with the red chart marker. Repeat the procedure with the other four missing numbers. Each time, after students have agreed on the number, put the number card in that slot and highlight it with the red marker. If students get restless, move quickly through these numbers.

When all five numbers have been guessed, ask students if they see any pattern in the red and black numbers.

After 30, which is the next number that should be red? Then which number? What about the numbers before 22?

Cover the black numbers with red inserts as students direct you. Continue the pattern up to about 40.

Take out the red inserts so only black numbers show for the final round. Remove five different numbers, but this time not in any pattern. Just hide five numbers that you think students will be able to identify. Ask students to tell you which number they think belongs in each blank space and to explain why they think it's that number.

Explain that students will be playing Missing Numbers again during Choice Time. They will play in pairs, either at the Hundred Number Wall Chart or using the plastic Hundred Number Boards with removable tiles.

Activity

Choice Time

Post the Choice Time list for the rest of today and most of the next two sessions. Since only three or four pairs at a time will be able to play Missing Numbers, you may want to establish an order in which pairs take turns with the available 100 charts.

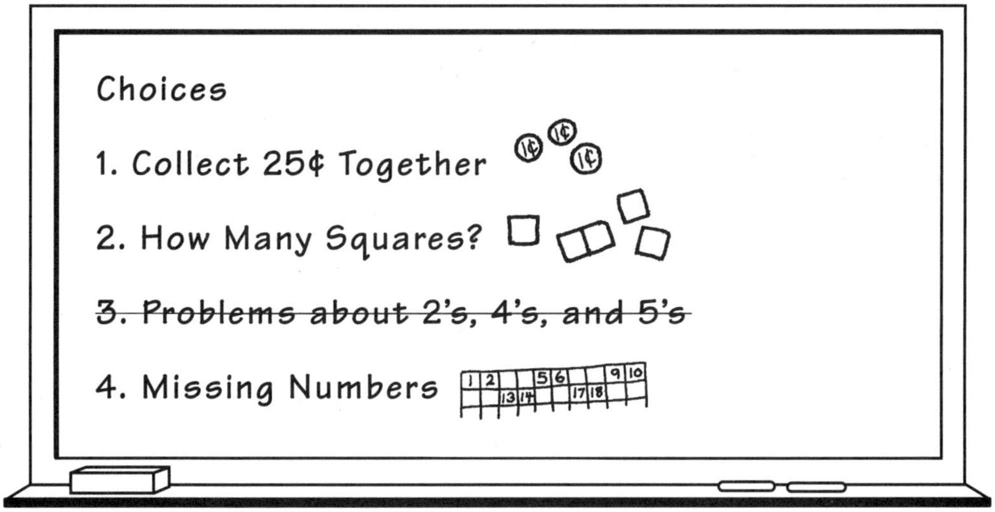

Choices

1. Collect 25¢ Together

2. How Many Squares?

~~3. Problems about 2's, 4's, and 5's~~

4. Missing Numbers

By the end of the three sessions, all students should have played Collect 25¢ Together and How Many Squares? If you have a small class or enough 100 charts with removable numbers, they may be able to complete their work with Missing Numbers as well. If your materials are limited and not everyone has a chance to try Missing Numbers during this Choice Time, plan to include the game in the next Choice Time (Sessions 10–12).

Interspersed in this Choice Time are two whole-group activities: Quick Images (to open Session 7), and sharing strategies for How Many Squares? (at the end of Session 8).

To review Choice 1: Collect 25¢ Together, see p. 72. Students who are ready for more challenge and who have the time can play a more advanced variation, such as Collect 50¢ Together. Again, if you notice that many students are starting to make trades among coins, a brief whole-class game gives you a chance to encourage those students to share their thinking about trades. At the same time, many first graders will not yet be ready to make trades and should understand that it's fine to continue playing with pennies only.

See p. 73 to review Choice 2: How Many Squares?

Choice 4: Missing Numbers

Materials: Hundred Number Wall Chart, with black numeral cards in place; plastic Hundred Number Boards with removable tiles (2–3 available)

While one student shuts his or her eyes, the other player in the pair removes and hides from five to ten number cards or tiles, either random numbers or numbers in a pattern. Players using the smaller boards must replace the missing numbers with blank tiles.

As each missing number is identified, the student who removed the numbers replaces the correct card or tile. When all the empty spaces are filled, the two players change roles. If materials are limited, you might specify that each student in the pair may take only two turns to remove tiles, so that more pairs get a chance.

For more challenge, students can remove more numbers. They can also remove several consecutive numbers and ask the other players to start by identifying numbers in the middle of the set of empty spaces. Some students might want to try removing all the numbers.

Observing the Students

See pp. 66–67 for guidelines on observing students' work on Choice 1: Collect 25¢ Together, and p. 74 for observing their work on Choice 2: How Many Squares?

Missing Numbers

- How do students decide what number is missing? Do they count from 1? Do they count from some other number? Do they use the numbers before and after the missing number? Do they use patterns in the number sequence? ("Every other number goes 2, 4, 6, 8, and this is 2 away from 36, so it's 38.") Do they have more than one way to find the missing number?
- Do students use the structure of the 100 chart to help them locate numbers? For example, do they know that a missing number is in the 40's because it's in the row that extends from 41 to 50? Do they know that a missing number must end in 3 because the others in the column do?
- What range of numbers are students comfortable with? numbers under 30? under 50? Do some students have particular difficulty with the teens, but find it easier to locate numbers in the 20's or 30's? Are some students comfortable with numbers anywhere on the chart?

Activity

Quick Image Squares

At the start of Session 7, present images from Quick Image Squares for two or three rounds of Quick Images. (See page 7 to review the procedure.) Some good choices would be images that suggest groupings of 2 or 4 (images I, J, and L) or groups of 5 (image K).

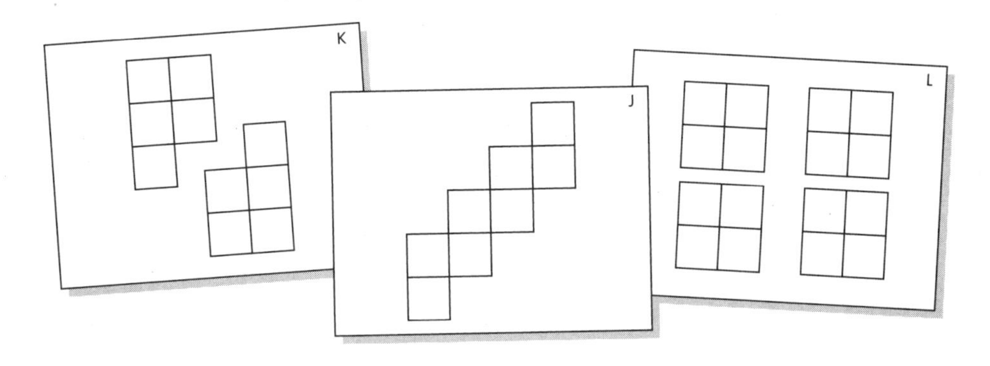

Activity

How We Counted

Take about 15 minutes at the end of Session 8 to discuss how students found the total number of squares in Set A. After each student explains a strategy, make a quick sketch of the arrangement of squares on the board or on chart paper and record the student's approach. Whenever possible, base your sketch on what the student recorded when doing the activity. Each time, ask students to raise their hands if their strategy is like the one just described; this gives more students a chance to participate in this brief discussion.

After This Choice Time Remember to collect the envelopes with Squares, Sets A and B and to recombine these, with your other squares, to make about 12 of Set C, as described in What to Plan Ahead of Time (p. 48).

Sessions 6, 7, and 8 Follow-Up

Coins Send home Student Sheet 14, Coins. In this problem, students find the total of several 2's and 5's, and they record their solution strategies.

❖ **Tip for the Linguistically Diverse Classroom** Read the sheet aloud, using actual coins to clarify the vocabulary as needed.

Session 9

Patterns of Fives and Tens

Materials

- Hundred Number Wall Chart, with red translucent chart markers
- Class sets of pattern blocks, interlocking cubes, materials such as Geoblocks
- Student Sheet 15 (1 per student and a few extras)
- 200 chart (optional, for challenge)
- Crayons or markers

What Happens

Class starts with Missing Numbers. Then, after doing Clapping Patterns, students represent a five-part pattern on a 100 chart and in another way (perhaps with drawings, cubes, or pattern blocks). Their work focuses on:

- making and describing repeating patterns
- representing the same pattern in different ways: through physical actions, concrete materials, drawings, and numbers
- exploring patterns in the number sequence

Note: If time permits, take a few minutes at the start or end of this session for students to share their strategies for solving the homework problem on Student Sheet 14, Coins.

Activity

Playing Missing Numbers Together

Gather students around the Hundred Number Wall Chart for a brief game of Missing Numbers. Remove the six number cards 20, 25, 30, 35, 40, and 45. Have the red translucent chart markers handy and use them to cover these numbers as you replace them.

As students tell you which numbers go in the blanks, ask them how they know. To involve more students, ask several students to offer their ideas on why a particular number goes in the blank. Or, give each student a square of paper and a pencil or crayon. As you point to each blank space, students who think they know the missing number can write it on their paper. When you ask all students who have written a number to hold it up, all those who are ready to participate can, while those who are not sure don't need to volunteer a number.

After all six numbers have been replaced and covered with the red chart markers, ask students what they notice about the red numbers and how they would extend the pattern in either direction.

After 45, which is the next number that should be red? Then which number? What about the numbers before 20?

Cover the black number cards with red inserts as students direct you, until you complete the fives pattern on the entire 100 chart.

Throughout this unit and the rest of the year, consider whole-class games of Missing Numbers as a good activity for spare moments.

Activity

Clapping Patterns

The activity Clapping Patterns is repeated throughout the grade 1 *Investigations* curriculum; students may remember it from the units *Mathematical Thinking in Grade 1* or *Building Number Sense.*

Gather students in a circle around you, so that everyone can easily see as you lead the patterns. Review (or introduce) Clapping Patterns, in which you and the class "act out" repeating patterns with a set of physical actions. Begin a simple pattern, such as knees-clap-clap (slap your knees, then clap twice) or knees-clap-tap (slap your knees with your hands, clap, then lightly tap your head). Keep a steady rhythm with one beat for each knee-slap, clap, or tap. After establishing the pattern, ask all students to join in as you repeat the pattern several times. Then do a few other simple patterns, perhaps some the students suggest.

The second part of the Clapping Patterns activity is to show the pattern in one or more other ways: with concrete materials, drawings, or numbers. If your students are new to the activity or if they have not done it recently, work with them on showing a pattern with only two different actions (such as knees-clap-clap), both with colored cubes and on the 100 chart. First ask students to describe the pattern to you in words.

What can you tell me about this pattern? Who has another idea?

Let's show this pattern with these red and blue cubes. I'm going to use *blue* for knees and *red* for clap. How could I make a cube tower to show our knees-clap-clap pattern?

If students don't understand this question, do the clapping pattern with them again. This time, while you are clapping, say the words "blue, red, red" as you act out the pattern. Then repeat the question.

Next, show this pattern with the first 30 or so numbers on the Hundred Number Wall Chart, using the red chart markers.

If the black numbers are knees and we make them red for claps, what should I do to represent knees-clap-clap on this 100 chart?

If necessary, demonstrate with the cards up to 6. Point to them as you chant "black, red, red, black, red, red." Then ask students to continue the pattern. Keep going up to 20 or 30, depending on students' attention and interest. Then ask students what they notice about the pattern on the 100 chart. For example, they might point out the three parts to the pattern, note that black makes a diagonal, or see that there are always either three or four blacks in each row.

Activity

A Pattern with Five Parts

For the rest of the session, students will work on showing a five-part pattern on paper 100 charts and with classroom materials (cubes, pattern blocks, and any other materials you have available) or drawings.

Present this pattern: knees-knees-clap-clap-tap. To give this pattern a nice rhythm, you can fit it into four beats, with clap-clap done quickly as one beat. The whole pattern, then, has the natural rhythm of the familiar rhyme "three, four, shut the door."

After students have done the pattern a few times, ask them to describe it in words. Then, explain that they will be showing this pattern in two ways. Each of them will color a 100 chart with three colors, choosing a different color for each kind of action (knees, clap, tap). They also show the pattern by drawing pictures or by using cubes, pattern blocks, or other such materials. Students who build their patterns with materials can trace or draw them to record their work.

To help everyone remember the pattern while they work, ask for suggestions for a way to record it on the board. Students might suggest you draw two knees, two pairs of hands clapping, and a hand on a head. Or, they might suggest you draw a sequence such as circle, circle, triangle, triangle, square. Choose one of the ways they suggest and record the pattern. Remind students that their own ways to show the pattern should be different from the one you have drawn on the board.

Observing the Students

- Can students show their pattern on the entire 100 chart? Do they show it accurately? Do they notice any repetitions? For example, do they see that the pattern repeats twice in each row, that coloring the pattern creates single-color columns, or that all the "tap" numbers end in 5 or 0? Do they use these repetitions to help them finish coloring the chart? Some students may find it helpful to act out the pattern as they color the chart.
- Do students represent their pattern accurately with materials?
- Can students explain what a particular number on the chart, or a particular part of their representation, shows about the pattern? For example, if you point to a row of the 100 chart they have colored, can they tell you which numbers stand for claps? If they have built a long train of interlocking cubes to show the pattern, can they identify the five-part unit that shows the basic pattern, knees-knees-clap-clap-tap?
- Can students predict the color of a square that they have not yet filled in on the 100 chart? For example, can students who have colored in numbers up to 30 predict the color of 50 without actually finding the color of each intervening square? Do they base their predictions on patterns in what they have done so far?

For More Challenge Students find additional ways to represent the knees-knees-clap-clap-tap pattern, or they create other patterns of their own and color them on the 100 chart. Students who seem ready to work with larger numbers could show their patterns on a 200 chart.

You may want to make a display of students' representations of the pattern. Encourage students to talk about ways that the various representations are similar to one another, and ways that they differ.

Sessions 10, 11, and 12

Twos, Fives, and Tens

Materials

- Dot cubes (2 per pair)
- Roll Tens Game Mats (5 of each, with extras available)
- Interlocking cubes (class set)
- Student Sheets 16–18 (1 of each per student, homework)
- Blank 100 charts (optional)
- Calculators (at least 6–8 for the class)
- Squares, Set C (12 sets)

What Happens

The game Roll Tens, in which players collect interlocking cubes and group them in rows of ten, is added to Choice Time, along with Exploring Calculators. A variation on How Many Squares? is used as an assessment. Students' work focuses on:

- developing strategies for organizing sets of objects so that they are easy to count and combine
- finding the total of several 2's, 5's, and 10's
- developing a sense of the size of the numbers up to 100
- recording strategies for counting and combining, using pictures, numbers, and words
- exploring calculators as a mathematical tool

Activity

Introducing Roll Tens

Roll Tens is a cooperative game in which pairs of students collect interlocking cubes in rows of 10 to fill a rectangular mat. Use the small Roll Tens 30 Mat to introduce the game, but emphasize the goal of *filling the mat,* rather than collecting 30 cubes, because some students may play with a larger mat and have a goal of collecting 50 or 100 cubes.

Gather students in a circle on the floor. Select a student volunteer to play a demonstration game with you. Have a supply of interlocking cubes, two dot cubes, and a Roll Tens 30 Mat (either the mat from p. 210, or one you have made to fit your cubes if they are not ¾ inch on a side).

My partner and I are going to show you a game called Roll Tens. The object of Roll Tens is to collect cubes to fill a game mat with rows of 10 cubes. Two players work together to fill the same mat.

You and your partner will take turns rolling two dot cubes and snapping together that number of cubes to make rows of ten. As you snap the cubes together in rows, lay them on the mat after each turn. The game is over when you fill up your mat.

Demonstrate a turn, rolling the dot cubes.

I just rolled a 3 and a 5. How many cubes should I take? What should I do with them? So, now we have 8 cubes. How many more do we need to complete a row of 10?

Collect the cubes, snap them together, and place them in a row along the top of your mat. If some students find it difficult to determine how many more cubes are needed to make up a row of 10, don't spend too much time on this question.

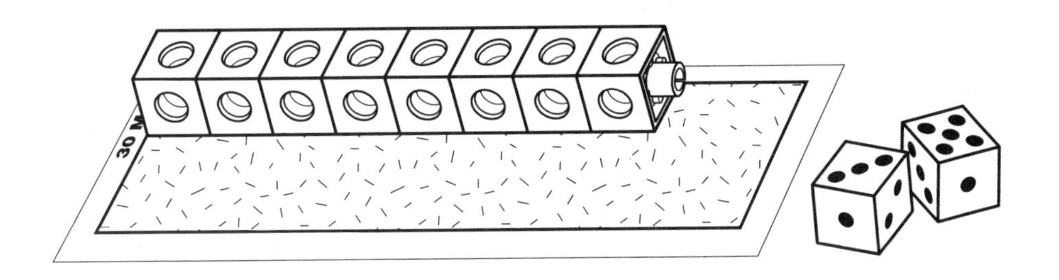

Then, ask your student partner to take a turn rolling the dot cubes. Again, involve the class in thinking about how many cubes to collect and what to do with them.

So, Brady rolled a 1 and a 4, and he took 5 cubes. After he uses them to finish our first row of 10, how many will be left to start the next row? How do you know? What's another way we could figure that out?

When the cubes have been snapped together and arranged on the mat, ask students about the cubes collected so far.

How many rows of 10 are on the mat so far? How many extra? How many do we have altogether?

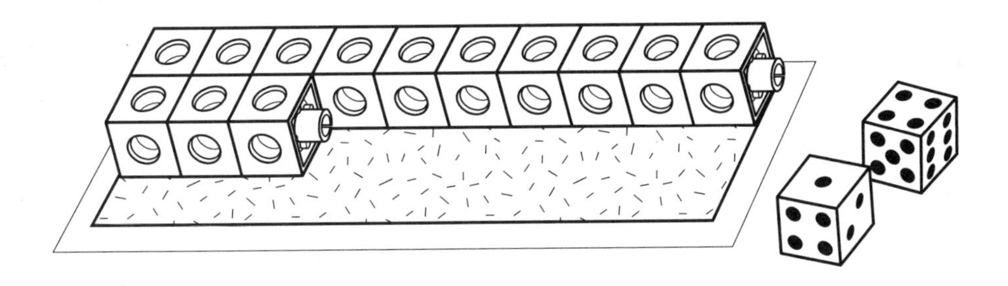

Continue to play for another couple of turns, involving the class in each turn, until you think students understand the game. Explain that playing Roll Tens in pairs will be one of their choices today and for the next two sessions.

Before You Begin Choice Time Take just a few minutes to introduce the two other choices, How Many Squares? and Exploring Calculators. Remind students of the activity How Many Squares? and explain that this time they'll use a new set of squares in envelopes labeled Set C. They will be working alone this time, and just as they did before, they will make a record of how they counted the squares.

Then explain that another choice is to explore calculators. The **Teacher Note**, Using the Calculator in First Grade (p. 163), offers reminders on appropriate uses of calculators in the first grade classroom. If students have not used calculators in a long while, you might take a few minutes for students to share some experiences they have had with calculators this year or ways they have seen calculators being used. Briefly review or introduce your class guidelines for caring for and storing the calculators.

Activity

Assessment

Choice Time

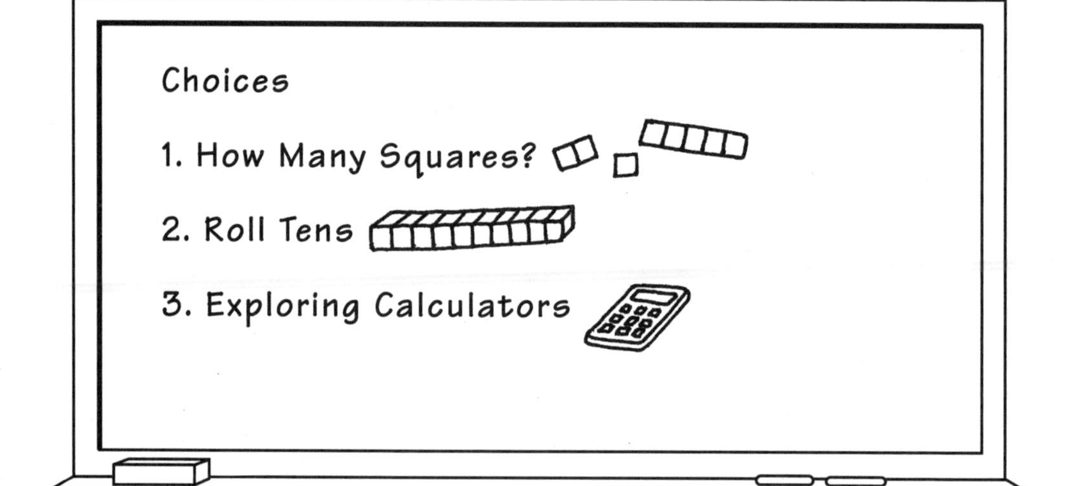

Assesment Note: The Choice Time actvities for the rest of Session 10 and most of Sessions 11 and 12 include the assesment activity How Many Squares? (Choice 1). Students should spend most of their time on How Many Squares? (the assessment) and Roll Tens. While is important for students to have opportunities to freely explore calculators throughout the year, they need not spend more than 10–15 minutes on calculators during this Choice Time.

You may find a wide range in how long students take to complete the assessment activity. Some may complete the work in 15–20 minutes; others may need a full session.

If you find that some of the class has finished all the choices while others are still working on the assessment, you can suggest additional choices, perhaps a variation of Collect 25¢ Together or Total of 10. Or, you might give them the homework on Student Sheets 16–18, What's Missing?, to start in class. Alternatively, you could bring Choice Time to a close when most students are ready and give a few students time outside of math class to finish the assessment.

Choice 1: How Many Squares? (Assessment)

Materials: Squares, Set C (12 sets, in envelopes)

Students count a set of squares containing single squares, pairs, and five-square strips. They make a record of how they counted the squares, including a picture of how the squares were arranged and an explanation (in numbers or words) of exactly what they did to find the total number of squares. Although students previously did this activity in pairs (with Squares, Sets A and B), emphasize that this time you want each student to find his or her own way to count the squares. As students finish, collect their written work to review as part of your assessment.

If you need to adjust the level of difficulty, remove squares from some sets and add squares to others. For more challenge, students can find and record more than one way to find the total number of squares.

Choice 2: Roll Tens

Materials: Dot cubes (2 per pair), interlocking cubes (in tubs of 50–100), Roll Tens Game Mats 30, 50, and 100 (5 of each, with extras available)

Student pairs take turns rolling two dot cubes. They collect that number of interlocking cubes, snapping them together into rows of ten. They continue collecting cubes until they have filled a game mat with rows of ten. They do not need to roll an exact number to complete the mat.

Depending on your class, you might offer a choice of mats and encourage students to find their own level of challenge. Or, everyone starts out with copies of the Roll Tens 30 Mat, and as pairs become ready for more challenge, they use one of the larger mats.

Choice 3: Exploring Calculators

Materials: Calculators (6–8 for the class)

This activity is open-ended. Your students may vary widely in what they notice about or do with calculator. Some may use calculators only to make numbers, some may use calculators to explore patterns, such as those created by repeatedly adding 2's, and some may make up and solve problems with calculators. Encourage students to share any discoveries they make about calculators.

Some students who can create and solve problems with calculators (or even without calculators) may have little understanding of what the operations they use mean, and little or no sense of whether the answer they obtain is reasonable. Do not insist that students work only with operations they understand well or with small numbers. Often when students are exploring calculators, they become fascinated by the power of being able to make large numbers or solve complicated arithmetic problems, even if they don't completely understand these.

Observing the Students

How Many Squares?

As you observe students working on this assessment activity, plan to jot down notes about how each student is approaching the task. With these notes and their written work, you can get a sense of their strategies for counting and combining 2's and 5's. **The Teacher Note**, Assessment: How Many Squares? (p. 95), includes sample student work.

- Do students organize the squares in some way to help them count and keep track, or do they count in a random order? Do they organize the squares by type (all the 2's together, all the 5's together)? Do they arrange the squares into equal groups, such as groups of four or five?
- How do students find the total number of squares? Do they count by 1's? Do they count by some other number? Do they count the squares appropriately—single squares as 1, pairs as 2, and so on? Do they use strategies involving number combinations?
- How do students record their work? Do their pictures of their arrangements show the correct number of squares? Do their pictures accurately show the way their squares are arranged? How clearly can they show their strategies? How are students using numbers to show what they counted?

Roll Tens

- How do students find the total of the two dot cubes? Do they count each dot? Do they count on from one of the numbers rolled? Do they use knowledge of number combinations?
- Do students know how many more cubes they need to complete a row of 10, or do they keep snapping together cubes and recounting until they have 10? Do students use knowledge of combinations of 10 to figure out how many more cubes they need to complete a row?
- If you ask students how many cubes they have collected on their mat so far, do they count all the cubes to find the total? Do they use tens and ones? ("Three rows of 10 is 30, and then three more is 33.") Are they confident that each row has 10 cubes, or do they count each row to check?
- How do students determine if they're finished? if they're almost finished? Do they focus only on whether they have filled the mat, or do they talk about whether or not they are close to the total?

You might call together any students having difficulty with Roll Tens to play a couple of rounds with you. Encourage them to count carefully and to share their approaches with one another.

If you have students ready for more challenge, see the extension activity How Many More Cubes? (p. 94).

Exploring Calculators

- Do students know how to turn the calculator on and off? Do they recognize that the digits they enter appear on the screen display? Do they know how to clear the screen?
- Are they familiar with the +, –, and = symbols? with any other symbols on the keyboard? Can they do any computations?
- What do students do with the calculator? Do they make numbers? explore patterns? create and solve computation problems?
- What numbers can they read? If a two-digit number appears on the screen display as a result of a computation, can they tell you what the number is? Do they know the names of any three-digit numbers that appear on the screen display?

Note: Students will likely encounter decimals during their calculator explorations. If some students want to know what "the little dot" means, ask them first for their ideas. Many first graders will not yet be ready to interpret the decimal portion of the number as a part of a whole number, and you can explain that they'll learn more about numbers with dots, or decimal points, in later years. For now, when they encounter a number like 5.326, they can think of it as "about 5," or "5 and a little more," or "5 and some extra."

What's Missing? (Optional)

If some students finish their work on all three choices and you decide to offer What's Missing? (Student Sheets 16–18) as a choice, look for the following as students work:

- How do students decide what number is missing? Do they count from 1? Do they count from another number? Do they use the numbers before and after the missing number? above and below the missing number? Do they use the structure of the 100 chart to help them?
- Do students have more than one strategy for finding the missing numbers? As students are working, ask them to "prove" that two or three of the numbers they have filled in are correct. Encourage them to find more than one way to show that the numbers are in the right place.
- Can students write numbers accurately? Do they reverse digits in some numbers?
- Are students comfortable finding numbers anywhere on the chart? Are there some number ranges students seem uncomfortable with? Do they find some rows or columns easier (or more difficult) than others?

Sessions 10, 11, and 12 Follow-Up

Homework

What's Missing? Students fill in Student Sheets 16–18, What's Missing? (A–C). Students may work on the sheets in any order, although Sheet 18 has more missing numbers and may be more challenging.

For easier sheets, fill in some of the numbers, or use blank 100 charts to make your own partly filled-in charts.

For more challenge, give students a blank 100 chart to fill in completely. Students who are ready to work with numbers greater than 100 can fill in a blank 200 chart (tape together two blank 100 charts).

Extensions

How Many More Cubes? Do this activity with either the Roll Tens 50 or Roll Tens 100 game mat. Snap together two rows of 10 cubes and another row of about 7 more cubes. Line them up on the mat, as if playing Roll Tens. Ask students how many cubes you have put on the mat and how they know. Gather a variety of strategies.

Then, ask a few questions about adding or taking away groups of ten.

What if we took away 10 of the cubes? How many would we have? How do you know? What if we added 10 more? 10 more after that? How many more would we need to fill the mat?

Gather several strategies for each question you ask. You can repeat the activity, starting with a different numbers of cubes each time.

Counting on the Calculator Demonstrate how to use the calculator to "count" by repeatedly adding 1. Some students enjoy using the calculator to count to 100, or even higher.

Collections of 100 For practice with counting objects to 100, students make collections of 100 things, such as paper clips, a poster of 100 stickers, or a long line of 100 interlocking cubes. Students can also look for things in the classroom that there are 100 (or more) of, such as books, markers, or children's drawings.

100 Children Prepare a set of cards numbered from 1 to 100. Work with other classes in your school to form a human row of 100 people. Each person takes a card, and they arrange themselves in order from 1 to 100.

Assessment: How Many Squares?

Teacher Note

The notes you take as students work on the assessment activity How Many Squares?, together with the written work you collect from students, can give you a sense of how they are organizing sets to make them easier to count and what strategies they have for finding the total of several 1's, 2's, and 5's. Following are examples of a range of approaches you might see in a first grade classroom.

Counting Each Square Brady removes the squares from the envelope and separates them on the table so none are overlapping. In the center of his work space, he puts together the single squares. Then he draws a careful copy of the arrangement of squares. He slowly counts the squares in his drawing, starting with his grouping in the center, and then moving around the drawing in what seems to be a random order. After he says a number, he records it on the picture of the corresponding square. He tells the teacher that he counted the squares in the drawing (rather than the actual squares) because once he writes on a square, he knows he has already counted it.

Brady's work

Brady demonstrates a competent first grade approach. He counts the squares in pairs and fives accurately, he has a way of keeping track of what he has counted, and he arrives at the correct total. The teacher has noticed that Brady uses strategies involving counting by 1's to solve a variety of problems, but she has not yet observed him counting by numbers other than 1. Many first graders are still becoming comfortable counting quantities in the 20's by 1's and are still developing techniques for keeping track of what they count. Over the next year or two, they will become more secure and confident counting by 1's and will begin making sense of counting by numbers other than 1.

Organizing the Squares in Groups and Counting Each Square When Susanna first takes the squares out of the envelope, she tells the teacher that she is going to put them in groups of four and count by 4's. She puts together the pairs of squares and two of the single squares to form three groups of four, and she puts the remaining squares in a row. She looks thoughtfully at the groups of four for a moment, points to one group and says "4," points to another group and says "6 ... no... 6, 8," and points to the third group, hesitates, and then quietly says "10." She pauses and then says she'll try a different way.

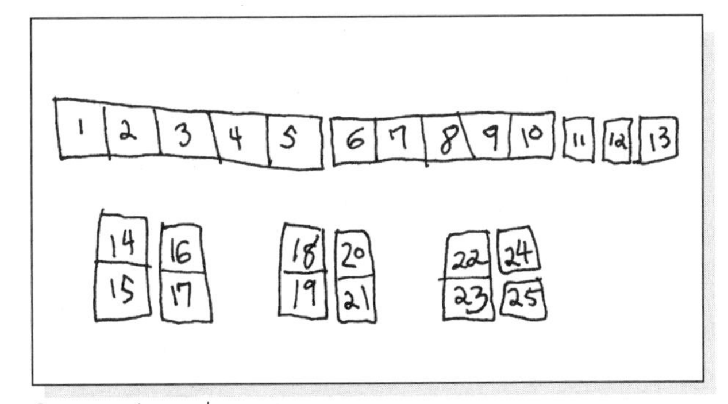

Susanna's work

This time, she begins by counting each square in the row, and then she counts each square in the groups, one group at a time. She copies her arrangement on paper and records the order in which she counted.

Like Brady, Susanna finds the total accurately by counting each square. However, she starts out by attempting to use a different approach: organizing some of the squares into groups of four and counting by 4's. Recently, the teacher has observed Susanna just starting to count by 2's and 4's for small amounts. Perhaps she abandoned this approach for this problem because she is not yet comfortable counting 12 objects by 4's, or perhaps she does not yet completely trust the totals she arrives at when counting by numbers other than 1. The teacher makes a note to give Susanna more experiences throughout the year that will help her explore different ways to organize and count sets. For example, she might ask Susanna to work on How Many Squares? or Collect 25¢ Together with Jacinta, who is beginning to count by 2's.

Organizing the Squares into Groups and Summing the Group Sizes Tamika begins by organizing the squares into several groups. She puts the two fives together, and puts the rest into three groups of four and a group of three. She copies her arrangement on paper and circles her groups (the two fives, two of the fours, and a four and three). She records the number in each group and the total number of squares enclosed by each circle. Then, she records an equation corresponding to each circled group. She works quickly and appears to be using knowledge of number combinations, rather than counting, to find the totals.

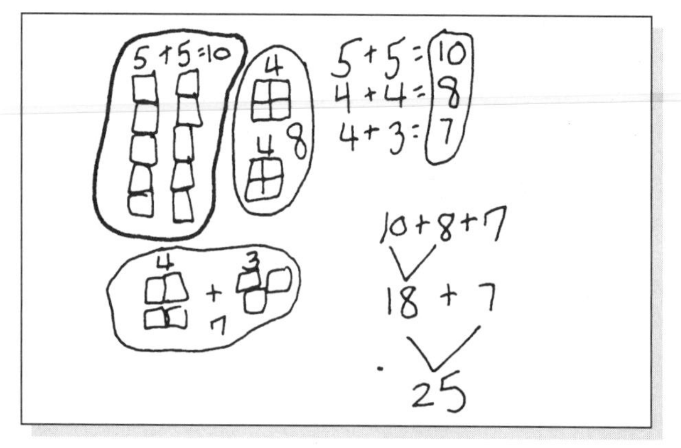

Tamika's work

Next, Tamika writes an expression in which the totals in her three equations are addends: 10 + 8 + 7. Below it, she writes another expression, in which she has combined the 10 and 8: 18 + 7. After hesitating for a moment, she counts up 7 from 18, keeping track on her fingers, and arrives at a total of 25.

When Tamika finishes, the teacher asks her to share her work with Jonah, who took a similar approach. Jonah organized the squares into groups of five, wrote 5 + 5 + 5 + 5 + 5, and found the sum by combining pairs of numbers.

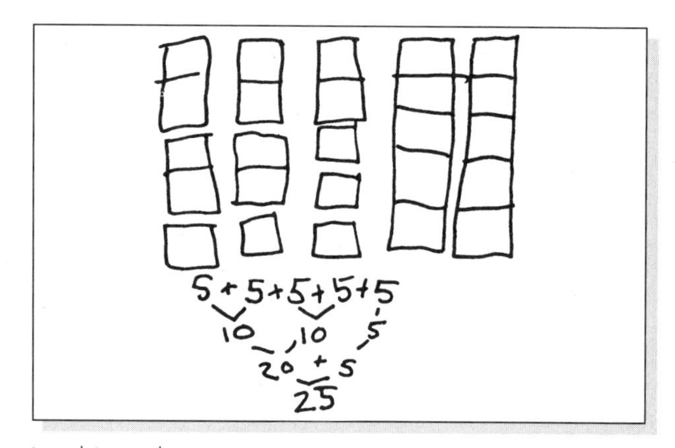

Jonah's work

Both Tamika and Jonah are able to combine small numbers by using number combinations they know. They arrange their squares into groupings that help them do this, and they don't need to count all the squares by 1's. Talking through their strategies and demonstrating ideas to one another can help them solidify their thinking and learn about new ways to organize and combine the squares.

Counting by Numbers Other Than 1 Michelle begins by sorting the squares: the five-strips in one corner, the pairs in another, and the singles in a third. She points to each strip of five in turn, saying "5, 10." Next, she counts the pairs, touching each as she counts "2, 4, 6, 8, 10." She says, "Two 10's are 20," then counts the single squares, again touching each square as she counts: "21, 22, 23, 24, 25."

Michelle's work

When the teacher asks Michelle if she could find the total in another way, she hesitates and then says that she could count each square. The teacher asks if there might be yet another way to count the squares. Michelle seems puzzled at first, but after moving the squares around for a moment, begins to organize them in groups of five.

Fernando also finds the total by organizing the squares into groups and counting them by the size of the group. When he takes the squares out of the envelope, he puts them in groups of five and counts by 5's. He quickly records his work, completing the problem in only a few minutes.

Fernando's work

The teacher asks Fernando if he could find the total in another way. He tells her he could do this in several ways: he could put the squares in groups of ten or he could do some tens and some twos or fours, but he couldn't put them all in groups of two or four, because he can't break up the fives.

Both Michelle and Fernando are comfortable counting the squares by numbers other than 1. While Michelle organizes the squares into groups and then counts the groups she has made, Fernando uses his strong understanding of number combinations and number relationships as he mentally plans ways he could organize, count, and combine the squares. Although he does not seem to need to work with the actual squares, he keeps in mind that numbers can be broken apart in some ways that the squares cannot be. For most students, the kind of flexible thinking about number that Fernando demonstrates develops gradually over the early elementary years.

One purpose of assessment is to get a sense of how individual students are working with specific ideas and concepts and to plan appropriate further experiences for those particular students. Another purpose can be to get a clearer sense of how the class as a whole is approaching these same ideas. After you have looked at each student's work, you might want to sort this set of papers by students who used a similar strategy. This can provide an overall picture of your class that might influence how you focus discussions or highlight specific strategies or ideas in future discussion.

For example, if many of your students are just beginning to find ways to organize squares into groups to make them easier to count, you might include How Many Squares? again in a future Choice Time, perhaps using a different combination of 1's, 2's, and 5's. If many are becoming comfortable counting small quantities by 2's, you might give them practice counting by 2's to larger numbers by repeating the activity of counting eyes, ears, hands, or feet in a large group or the whole class.

Session 13 (Excursion)

Counting by Kangaroos

Materials

- *Counting by Kangaroos*
- Counters (available)
- Drawing paper
- Crayons or markers

What Happens

Students listen to the story *Counting by Kangaroos,* by Joy N. Hulme, which illustrates finding the total of several equal amounts. Students then find the total for three groups of 11, 12, or 13 animals. Their work focuses on:

- finding the total of three groups of 11, 12, or 13
- recording strategies for counting and combining, using pictures, numbers, and words

Activity

Counting by Kangaroos

This excursion session is built around the book *Counting by Kangaroos,* by Joy N. Hulme. This story, in rhyme, tells about three kangaroos that come to visit a pair of young human friends. In their pouches, the three kangaroos are carrying groups of other Australian animals. Each group is a different size, from 3 to 10 (3 squirrel gliders, 4 koalas, 5 bandicoots, and so on up to 10 wallaby joeys). Each new page or facing spread of the book reveals the latest groups of animals to emerge from the three pouches, along with a rhyme about them. Each rhyme concludes with a count of the animals that have just emerged from the three pouches, like this one: "Squirrel gliders in a line. Count them—3, then 6, then **9**." The story continues similarly with 4, 8, **12**; 5, 10, **15**; and so on.

Read the book with the class. If your class is small and students can easily see the pictures, ask them to count some of the animals on the pages along with you.

When you have finished, pose a problem based on the book. Choose one of the scenarios near the start of the book, such as the three squirrel gliders leaping from the pouches. Reread the first part of the rhyme, but stop before you reach the part in which the count of the animals is given. Put the book down, so that students can't see the numbers on the page or count the animals.

So, each of the three kangaroos had three squirrel gliders in her pouch. How many squirrel gliders were there? How do you know? Did anyone find the number of squirrel gliders in a different way?

Students may use counters or paper and pencil to help them. If some students noticed the numbers on the page or counted the animals, ask them not to tell the total, but emphasize that knowing the total itself isn't what's important—you are looking for different ways they could *figure out* the total.

Ask for a few students to share their ideas on the total and to explain their strategies. Hold up the book again, and then reread the rhyme, this time completing it. To double-check, students can count the animals in the picture along with you.

Pose a similar problem based on a different grouping early in the book, for example, the five bandicoots in each of the three pouches.

Activity

Finding the Total in Three Groups

As an extension to the story, each student decides on another kind of animal to emerge from the three kangaroos' pouches. Ask them to choose a group size just over 10, such as 11, 12, or 13. Students find the total number of their chosen animal (the total of three groups of 11, or 12, or 13) and show how they found the solution.

If time remains at the end of the session, students can share their work with the class.

Session 13 Follow-Up

How Many Animals in All? Flip through the book *Counting by Kangaroos* with the class and list together the successive groups of animals that came out of each pouch: 3 squirrel gliders, 4 koalas, 5 bandicoots, and so on. Students find the total number of animals that each kangaroo had in her pouch. For extra challenge, students can find the total number of animals in all three pouches.

Addition and Subtraction

What Happens

Session 1: Combining Situations Students extend their understanding of combining problems, in which they find the total of two amounts. They record and share their solution strategies.

Session 2: Separating Situations Students extend their understanding of separating problems, in which they find the result when one quantity is removed from another. They record and share their solution strategies.

Sessions 3, 4, and 5: Five-in-a-Row and Story Problems Students play a challenging version of Five-in-a-Row, a game introduced in an earlier unit that provides practice with single-digit addition pairs, this time with sums up to 20. They play this game and solve story problems for Choice Time. At the end of Choice Time, students share strategies for some of the story problems they solved.

Sessions 6, 7, and 8: Tens Go Fish In the game Tens Go Fish, students make combinations of ten with two addends. This game is added to the previous choices (Five-in-a-Row and story problems) as Choice Time continues for three more sessions. A whole-group round of Quick Images starts Session 7, and during the last half of Session 8, students share strategies for some of the story problems they solved.

Session 9: Combining with Unknown Change Students work with another type of story problem, "combining with unknown change" (a total and one of the amounts are given, and the second amount must be found). Again, they record and share their solution strategies. Then, for homework, they write their own story problems to match a given addition expression.

Sessions 10, 11, and 12: Addition and Subtraction In the unit's final Choice Time, students work on Total of 20, (a variation of a game from Investigation 1), Tens Go Fish, and story problems. During the last half of Session 12, students share strategies for solving story problems.

Session 13: Solving Story Problems As an assessment, students solve a variety of combining and separating story problems and record their solution strategies.

Routines Refer to the section About Classroom Routines (pp. 166–173) for suggestions on integrating into the school day regular practice of mathematical skills in counting, exploring data, and understanding time and changes.

Mathematical Emphasis

- Visualizing combining and separating problem situations
- Developing strategies for solving combining and separating story problems
- Recording strategies for solving combining and separating story problems, using pictures, numbers, words, and equations
- Becoming familiar with combinations of 10 and 20
- Reasoning about combinations of 10
- Increasing familiarity with single-digit addition pairs

SET A 1. Ken found 12 white shells at the beach.
He found 6 brown shells.
How many shells did he find?

SET A 7. I see 4 children and 2 dogs with muddy feet.
How many muddy feet do I see?

What to Plan Ahead of Time

Materials

- Number Cards: 1 deck per pair (Sessions 3–8, 10–12)
- Overhead projector (Sessions 3–8)
- Quick Image transparencies, Squares or Dot Addition Cards (Sessions 6–8)
- Interlocking cubes (Sessions 9–13)
- Paste or glue sticks (Sessions 3–8, 10–13)
- Counters, such as buttons, bread tabs, or pennies: at least 40 per pair (available)
- Unlined paper (available)
- Chart paper or newsprint (18 by 24 inches): 15–20 sheets (available)
- Envelopes for story problems (at least 21)

Other Preparation

- Duplicate student sheets and teaching resources, located at the end of this unit. If you have Student Activity Booklets, copy only items marked with an asterisk.

  For Session 1

  Student Sheet 19, At the Beach (p. 212): 1 per student, homework

  For Session 2

  Student Sheet 20, Clay Animals (p. 213): 1 per student, homework

  For Sessions 3, 4, and 5

  Story Problems, Set A (p. 220): 1 per student and 1 extra set. Cut apart and sort into seven envelopes. Paste one copy of the problem on the envelope for identification.

  Student Sheet 21, Five-in-a-Row with Three Cards (p. 214): 1 per student, homework

  Student Sheets 22–24, Five-in-a-Row Boards A–C (pp. 215–217): 1 of each per pair and a few extras (class), plus 1 of each per student, homework. Prepare a transparency of Board A.*

  For Sessions 6, 7, and 8

  Story Problems, Set B (p. 221): 1 per student and 1 extra set. Prepare like Set A.

  Student Sheet 25, Tens Go Fish (p. 218): 1 per student, homework

  For Session 9

  Student Sheet 26, Write Your Own Story Problem (p. 219): 1 per student, homework

  For Sessions 10, 11, and 12

  Story Problems, Set C (p. 222): 1 per student and 1 extra set. Prepare like Set A.

  Story Problems, Set D* (Challenges) (p. 223): enough for about half the class (optional)

  For Session 13

  Story Problems, Set E (p. 224): 1 per student. Cut apart and clip in five sets.

- On chart paper, write a combining story problem and a separating story problem for use in Sessions 1 and 2. Read through these sessions for information on choosing appropriate problems. On a third sheet, write a story problem that involves combining with unknown change for use in Session 9 (see p. 138).
- If your Number Cards are duplicated on paper, make "card holders" for the game Tens Go Fish. Fold a letter-size sheet in half the long way. With the fold at the bottom edge, fold up about 1 inch and staple ends to form a pocket. See illustration p. 132.

Session 1

Combining Situations

Materials

- Combining problem on chart paper
- Unlined paper
- Student Sheet 19 (1 per student, homework)

What Happens

Students extend their understanding of combining problems, in which they find the total of two amounts. They record and share their solution strategies. Their work focuses on:

- visualizing what happens in combining situations
- understanding that when two amounts are combined, the result is more than the initial amounts
- developing strategies for solving combining story problems
- recording strategies for solving combining story problems, using pictures, numbers, words, and equations

Activity

Making Sense of Combining

Note: If your students have worked in the unit *Building Number Sense,* they will have had experience with both combining and separating problems. First graders need many opportunities to develop their understanding of story problems and to learn ways of recording their thinking clearly.

In this first activity, students solve story problems about combining by visualizing the amounts and the result. See the **Teacher Note,** Types of Story Problems (p. 108), for a discussion of the problem types students will be working with. Like the story problems in Investigation 2, the problems in this session involve finding a total. In Investigation 2, students found the total of several equal sets or groups: the number of hands in a group of people, or the number of wheels on several cars. In this session, the problems involve a sequence of actions in which two quantities are combined. Students need to figure out what is happening in the story: What does each amount represent? Are amounts being combined or separated? Will the result be more or less than the initial amount?

Interpreting story problems that describe a sequence of actions can be challenging for first graders. In this whole-class activity, the numbers in the problems are deliberately kept small so that students can work mentally and can focus on the meaning of the story problem. In the next activity, Recording Combining Strategies, students will continue to work on making sense of combining story problems and they will record their solution strategies.

A Combining Problem Tell a story problem like the one that follows. (This is *not* the problem you have written on chart paper.) If you make up a problem with a more familiar or timely context for your students, keep the same numbers and basic structure. See the **Teacher Note,** Creating Your Own Story Problems (p. 164), for other considerations.

❖ **Tip for the Linguistically Diverse Classroom** Make story problems comprehensible to all students by acting them out and/or drawing quick sketches of important words (*birds, branch*).

An important aspect of solving story problems is being able to recognize the sequence of actions in the story. Ask students to try to see the story in their minds as you tell it. Some may want to close their eyes to help them concentrate.

The other day, I was watching some birds at the park. I counted 5 on one tree branch, then 2 more birds flew over and landed on the same branch.

Who can tell me what happened in the story?

Ask students to tell what they remember: What happened first? Then what happened? Ask for the story from several students, even when one tells it correctly.

Some students may anticipate a final question, such as "How many birds were on the branch?" Remind them that for now, you're interested only in what they remember about the story and how they're thinking about it; you're not looking for an answer.

Next, ask students whether there were *more* or *less* than five birds at the end of the story, and how they know. Keep the focus of the discussion on *how they know.* Students might say they saw in their heads more birds flying in to join the first group; they might model the problem on their fingers; or they might explain that 5 and 2 more has to be more than 5.

Present another story situation, still keeping the numbers small (adding on 2 or 3 to a familiar number). Again, ask students to visualize what is happening as you tell the story.

Later that day in the park, I saw some squirrels on a log. I counted 6, and then another 3 scampered up on the same log.

After students retell the story in their own words, discuss whether there are *more* or *less* than 6 squirrels at the end of the story, and how they know.

Activity

Recording Combining Strategies

Now turn to the problem you prepared on chart paper, telling the story to the class before you show the written form. If you make up your own problem, keep the same numbers and basic structure as the library book problem included here. Even though some students may be comfortable finding sums of larger numbers, small numbers help all students focus on recording how they found their solutions. Later in this investigation, you can adjust the numbers in the problems for students ready to work with larger numbers.

❖ **Tip for the Linguistically Diverse Classroom** Continue acting out these stories, linking words like *books* and *book bag* to actual objects, if possible.

The other day I was getting some books from the library. I checked out 8 books and put them in my book bag. As I was leaving the library, I saw 5 more books that I really wanted to read to you, so I checked those out, and put them in my book bag, too.

Now who can tell me what happened in the story?

As before, keep the emphasis on retelling the story rather than anticipating a question and answer. Listen to two or three students put the story in their own words. When you are satisfied that students have a good grasp of the story, show them the chart paper version of the story.

> At the library, I took out 8 books. Then, I took out 5 more books. How many books did I take out?

Distribute unlined paper. Explain that students not only find the answer but also show *how they found it.* For example, if they modeled the problem with cubes and then counted the cubes, they can draw a picture of the cubes and the numbers they said as they counted them. If they counted on their fingers, they can draw a picture of their hands and the numbers they said as they counted. Students work individually, but they may discuss their strategies with each other.

Observing the Students

- How do students approach the problem? Do they model it with objects or pictures? Do they work mentally or count to themselves? Do they record numbers or equations to help them find the solution?

  Some students may immediately pull the numbers 8 and 5 out of the story problem, but not know where to go from there. You can encourage them to find a model that works for them—cubes, or pictures, or fingers to represent the books.

- What strategies are students using? Do they count out all the quantities in the problem? Do they count on from one number? Do they use knowledge of number combinations? Do they use number combinations they know to find ones they don't know?
- Can students explain their strategies? How clearly can they record them? Do they use pictures? words? How are they using numbers or equations as part of their recording?

Students who say they found the answer by counting may need help recording this strategy. You can suggest that they use numbers or simply tell in words what they did.

If some students record 8 + 5 = 13 and say they "just know" the answer, encourage them to find a way to prove it. The emphasis in this activity is on explaining and recording thinking. All students, even those who have already committed 8 + 5 to memory, need to show a strategy for solving this problem. The work below gives some examples of how students might record their strategies.

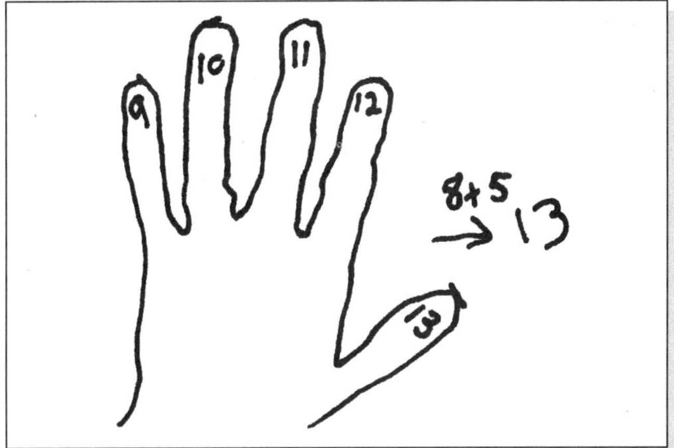

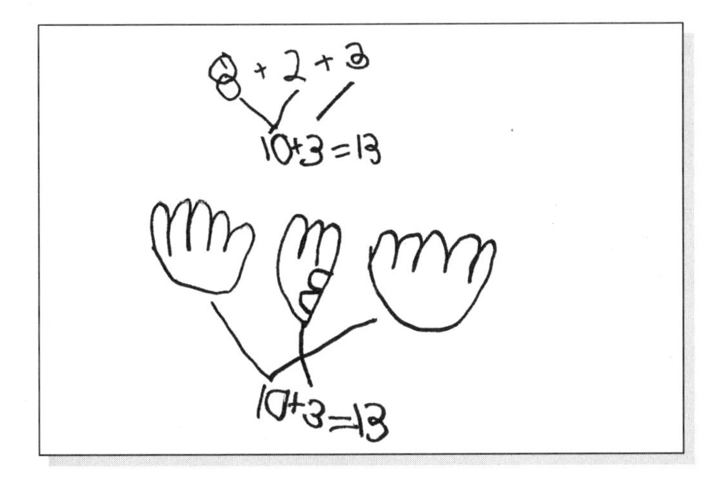

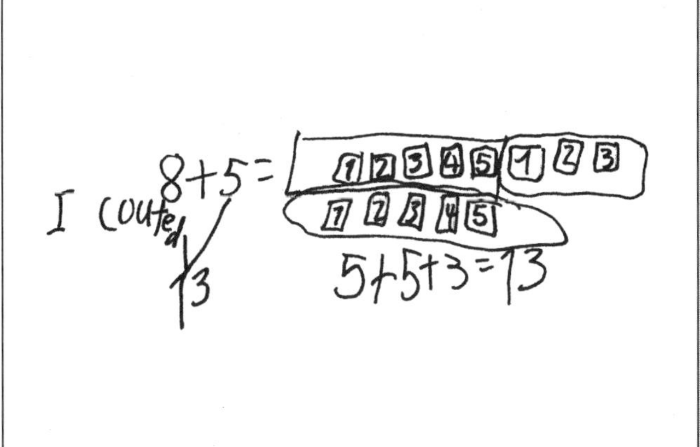

Throughout the investigation, occasionally ask students to prove some of the larger combinations they know. This can give them practice explaining and recording their thinking, and help them think of different strategies for solving a problem.

As you observe students at work, notice the variety of approaches and look for particular students you will ask to share their work with the whole class at the end of the session.

If you have students who solve this problem easily and record their strategies well, you can ask them to show a second way to find the answer or to solve a related problem with larger numbers. For example:

> So, I had 13 library books. Then, this morning I checked out another 8. Now how many do I have?

Activity

Sharing Combining Strategies

Gather the group together to share a few strategies and recording methods. Have counters available for those who want to demonstrate their strategies. Record each different strategy on the board or on chart paper, as you did when students shared problem-solving strategies in Investigation 2. Whenever possible, base your recording on the way the student recorded.

Use equations if students themselves have used them or if they are a natural way to represent a strategy someone describes. If addition equations do not come up during this discussion, model them for the class at the end of this activity.

After recording five or six different strategies, or when students are becoming less attentive, ask students to decide which approach you've recorded is closest to their own. Take a show of hands for each approach.

If you have not yet recorded the problem with addition equations, write 8 + 5 = 13. If necessary, review the meaning of the symbols + and =. See the **Teacher Note,** Introducing Notation, page 162.

On the following page are several approaches to the library book problem and the way one teacher recorded them as students described their strategies to the class.

William: I drew 8 books, then I drew 5 books, then I counted them all.

Kaneisha: First there were 8 books, then I counted 5 more.

Luis: I used fingers to count 8 and 5.

Leah: I took 2 from the 5 and added it to 8 to make 10, then there's 3 more from the 5, so I added it to 10.

Nathan: I knew 3 and 5 is 8, so I did 5 and 5 is 10. Then I took the 3 that was left and I counted up 3 from 10.

1. 8 books ☐☐☐☐☐☐☐☐ 1 2 3 4 5 6 7 8 — 5 books ☐☐☐☐☐ 9 10 11 12 13

2. 8 ● ● ● ● ● (1 2 3 4 5) 9 10 11 12 13

3. 1 2 3 4 5 6 7 8 9 10 11 12 13 (fingers)

4. $8 + \overbrace{2+3}^{5}$
$8+2=10 \quad 10+3=13$

5. $3+5=8$
$5+5=10$ ● ● ● 11 12 13

Session 1 Follow-Up

At the Beach Students solve another combining problem at home. Send home Student Sheet 19, At the Beach. Remind them to show clearly how they found their solution, just as they did in class.

❖ **Tip for the Linguistically Diverse Classroom** Read the problem aloud. Students may add their own rebus drawings for key vocabulary *(children, beach, more).*

Teacher Note — *Types of Story Problems*

The two most familiar types of addition and subtraction story problems are about *combining* two quantities to find a total, and about *separating,* or removing one quantity from another. Within these two problem types, there are a variety of problem structures, defined by what information is given and what is to be figured out. The structure students work with in this unit are described below.

Combining In *joining* or *combining* problems, two or more quantities are combined to form another quantity. These include traditional addition situations:

> Ted had 6 marbles. Sophia gave him 4 more. How many does Ted have now?

In a situation like this, the quantities to be combined are known, and the total is to be found. These problems with an *unknown outcome* are probably the most familiar type for your students.

In another type of combining problem, with *unknown change,* the total amount and one of the quantities are known, and the second quantity must be found:

> Ted had 6 marbles. Sophia gave him some. Now Ted has 10 marbles. How many did Sophia give him?

This structure is more challenging for young students than the more typical problem with an unknown outcome. Some students might solve this problem by a method based on subtraction, such as counting back from 10 to 6; others might use a method based on addition, such as counting up from 6 to 10. However it is solved, the structure is considered combining: Two quantities are combined to create a third. In this case the outcome is known, but the change that occurs is not known.

Separating In separating problems, one quantity is removed from another, resulting in a portion of the original quantity. These include the familiar "take away" situation:

> Ted had 10 marbles. He gave 4 marbles to Sophia. How many does he have now?

This is a separating problem with an *unknown outcome:* we start with one quantity, change it by removing one or more parts, and find the remaining quantity. As with combining problems, we can alter the structure of the problem by changing what information is given and what we have to find out. In this way we form a separating problem with *unknown change:*

> Ted had 10 marbles. He gave some to Sophia. Now he has 6. How many did he give away?

Separating problems with an unknown change, like combining problems with an unknown change, are more difficult for young students to make sense of.

Students' work with story problems in the *Investigations* curriculum at grade 1 focuses on making sense of and developing strategies for solving combining and separating problems with unknown outcomes. In this unit, students also become familiar with combining with unknown change, and those ready for more challenge encounter separating with unknown change.

Some researchers classify the addition and subtraction problem types described here in slightly different ways. However, this information is all you need to begin recognizing the different structures you will use in providing appropriate problems for your students.

Session 2

Separating Situations

What Happens

Students extend their understanding of separating problems, in which they find the result when one quantity is removed from another. They record and share their solution strategies. Their work focuses on:

- visualizing what happens in separating situations
- understanding that when one amount is removed from another, the result is less than the initial amount
- developing strategies for solving separating story problems
- recording strategies for solving separating story problems, using pictures, numbers, words, and equations

Materials

- Separating problem, on chart paper
- Unlined paper
- Student Sheet 20 (1 per student, homework)

Activity

Making Sense of Separating

This session repeats the steps of Session 1, but shifts the focus to separating problems. Students visualize what happens when one amount is subtracted from another: Do you have more or less than when you started? Again, the numbers are kept small so that students can work mentally and focus on the meaning of the story problem.

To prepare students for working with the problem you've written on chart paper, first tell a story problem like the one that follows. Ask students to listen carefully and try to see in their minds what is happening in the story. Closing their eyes can help some students concentrate.

❖ **Tip for the Linguistically Diverse Classroom** As before, make stories comprehensible by acting them out and making quick sketches as a visual clue to important words.

Here's another story about something that happened when I was in the park. I was walking along a path, and a saw a little pile of leaves on the ground in front of me. There were 8 leaves. Then, the wind blew away 2 of the leaves.

Ask a few students to tell the story back to you. Then pose the following:

When I first saw the pile, there were 8 leaves in it. Were there *more* or *less* than 8 leaves in the pile at the end of the story? How do you know?

If you think students need more practice making sense of separating, try one more story problem with the whole group.

So, now there were 6 leaves in the pile. But the wind was still blowing, and it blew 3 more leaves away.

Again, volunteers retell the story, say whether there would be more or less than 6 at the end of the story, and tell how they know.

Activity

Recording Separating Strategies

Tell the story of the problem you prepared on chart paper before showing it to the class.

I have one more story to tell you about my trip to the park. It was a beautiful day, and I saw lots of children playing there. When I got there, I saw 14 children. Then 5 of them left.

Again, ask two or three students to put the story in their own words. Keep the emphasis on retelling the story and discourage them from anticipating a question and answer. When you are satisfied that students have a good grasp of the story, show them the chart paper with the problem they are to solve.

> 14 children were playing in the park. Then 5 children left. How many children were still in the park?

Distribute unlined paper. Emphasize that students need to record their work in a way that would help someone else understand how they found their solutions.

Most students will probably use a strategy based on subtraction, such as "taking away" 5 from 14, or counting down 5 from 14. Some students may use strategies based on addition and solve the problem by counting up from 5. See the **Teacher Note,** Three Approaches to Story Problems (p. 113), for some typical student approaches to a similar separating problem. The **Teacher Note,** The Relationship Between Addition and Subtraction (p. 116), discusses some important connections between these two operations.

If some students solve this problem easily and record their strategies well, they might show a second way to find the answer or solve a related problem with larger numbers:

> Later that day, 18 children were playing in the park. It started to rain a little bit, and 14 of the children went home. How many were still in the park?

Activity

Sharing Separating Strategies

Gather the group together to share a few of their strategies and recording methods. Have counters available. Record each different strategy on the board or on chart paper. As before, base your recording on the student's own approach.

Watch for (but do not expect) the use of equations. Although some students may be familiar with subtraction notation from their work in the unit *Building Number Sense,* they may find it more natural to represent their strategies in other ways. If some students use notation incorrectly ($5 - 14 = 9$ instead of $14 - 5 = 9$), model the correct form on the board. The **Teacher Note,** Introducing Notation (p. 162), suggests how to handle the incorrect use of notation. If no one uses a subtraction equation and you don't find a natural way to use one as you record students' strategies, plan to model subtraction equations at the end of this activity.

Here are the ways one teacher recorded some students' different approaches:

Yanni: I drew 14 people, then I crossed out 5 and counted what was left.

Libby: I put out 14 cubes, and then I took away 5. I counted how many were left.

Jamaar: I counted 5 back from 14.

Shavonne: I counted up from 5 and kept track of how many numbers I counted.

Nadia: I knew that 14 take away 4 is 10, and then take away 1 more is 9.

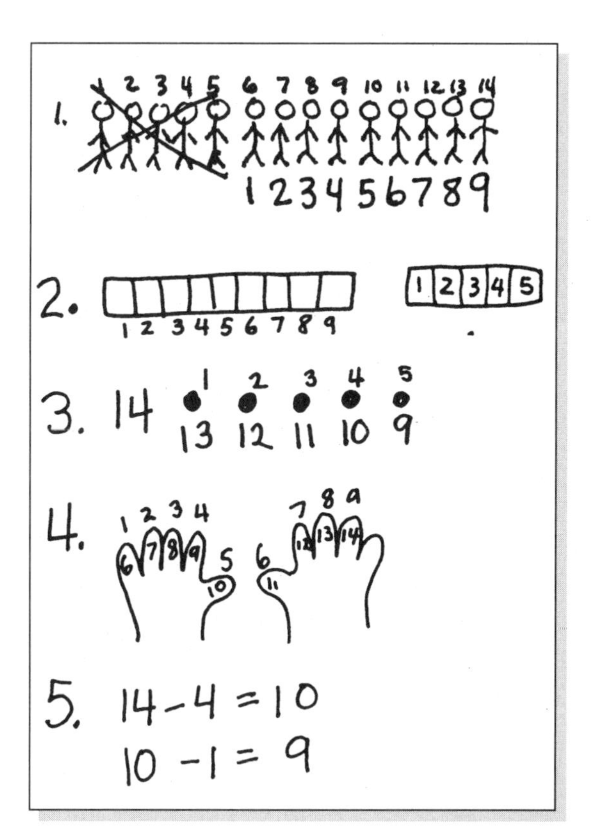

After recording several approaches, ask students to look at their papers and decide which of the listed strategies is most like their own.

If some students used "take away" or counting back strategies and others used adding on strategies, draw attention to these different ways of thinking about the problem:

Chris counted up to solve this problem, but Andre counted down. How can they both work?

If you have not yet used subtraction equations, write 14 – 5 = 9 and ask students what they think the minus (–) symbol means. Use the terms *minus* and *subtract* in connection with this symbol so that students become familiar with both. While *take away* is not a mathematical term, its use by students need not be an issue.

At the end of the session, collect both last night's homework and students' work from today. Review it to see if students are representing their strategies clearly, with words, pictures, and numbers.

Session 2 Follow-Up

Clay Animals Send home Student Sheet 20, Clay Animals, to give students a separating problem to solve at home. Remind them that they need to show how they found their solution, just as they have been doing in class.

❖ **Tip for the Linguistically Diverse Classroom** Read the problem aloud. Students may add their own rebus drawings for key vocabulary *(clay animals, gave, friends).*

Three Approaches to Story Problems

Teacher Note

Students commonly take one of three approaches to solving a story problem: direct modeling, counting strategies, and numerical reasoning. Each approach is described below and illustrated with examples of student work on this separating problem:

> Last night I picked up 12 pencils from the floor. I put 4 of the pencils in the pencil box. How many pencils did I have left in my hand?

Direct Modeling

When young students encounter story problem situations, they usually model the actions in the problem step by step in order to solve it.

Students who are using a direct modeling strategy might count out 12 cubes, then take 4 of them away to represent the 4 that were put in the pencil box, then count the number of remaining cubes.

Leah drew twelve pencils, crossed out four, and then counted the remaining set.

Fernando counted out twelve cubes, took away four, and counted the remaining cubes.

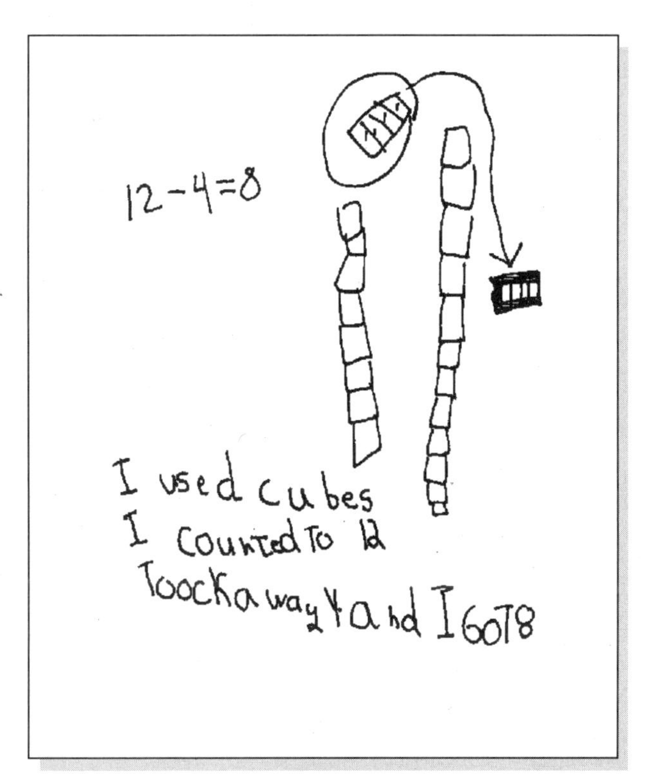

As students gain skill in visualizing problem situations and begin to develop a repertoire of number relationships they know, they gradually develop other strategies based on counting and on numerical reasoning. These strategies require visualizing all the quantities of the problem and their relationships, recognizing which quantities you know and which you need to find.

Counting On or Counting Down

Some students, who perhaps feel more confident visualizing the problem mentally, use strategies that involve counting on or counting down.

Donte counted on his fingers. To get 12, he explained that he used both hands and visualized two "imaginary fingers." He counted down from 12, first counting down 2 in his head, using his imaginary fingers, then counting down 2 more on his actual fingers, and got 8.

Continued on next page

Teacher Note

continued

Donte recorded his counting down strategy on paper like this:

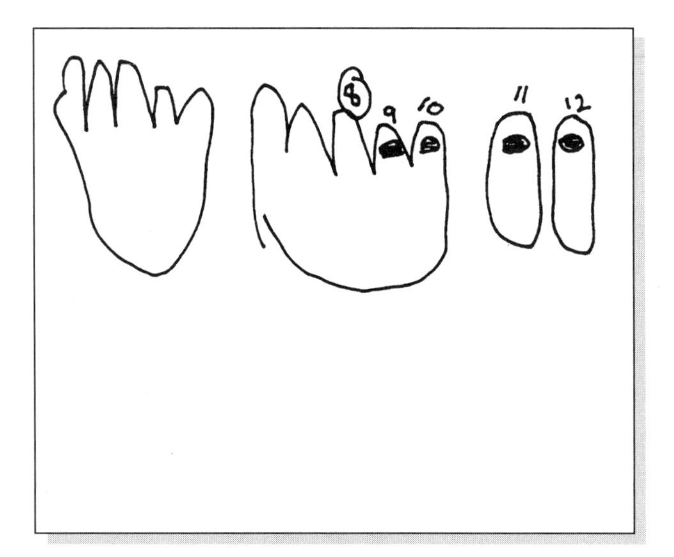

Kristi Ann used the class number line. She started at 12, and counted back 4.

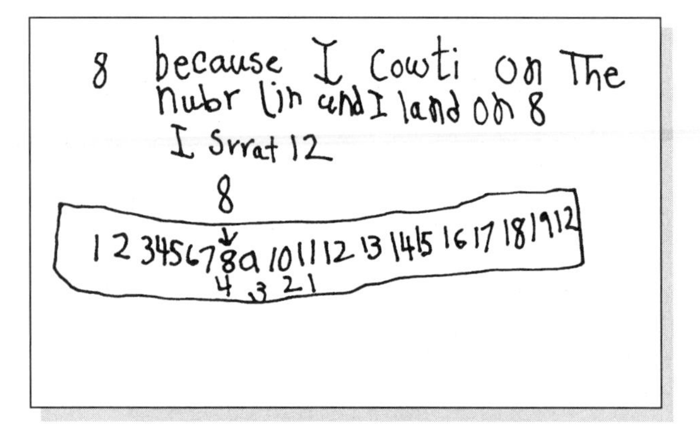

Andre counted down 4 from 12 in his head.

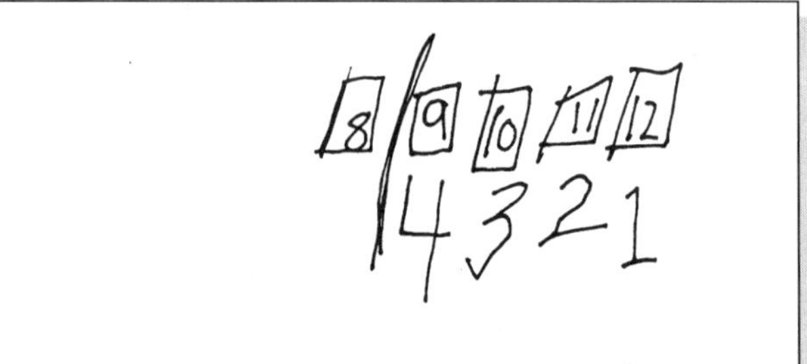

While these three students' methods somewhat resemble the methods of the students that directly modeled the action in the problem, there is an important difference: None of these students had to construct the 12 from the beginning, by 1's. Donte quickly made the 12 out of larger chunks (5 + 5 + 2), while Kristi Ann and Andre simply started with 12. Counting back for subtraction requires a complex double-counting method. These students must simultaneously keep track of the numbers they are counting down (11, 10, 9, 8) and the number of numbers counted (1, 2, 3, 4).

Iris used a different counting strategy. She counted out four cubes in a row, to represent the four pencils taken away. She continued putting cubes in a second row, counting on from four until she had a total of 12. Then, she counted the number of cubes in the second row. Iris was able to transform the problem into a different structure, then count on to find the solution: 4 + ___ = 12.

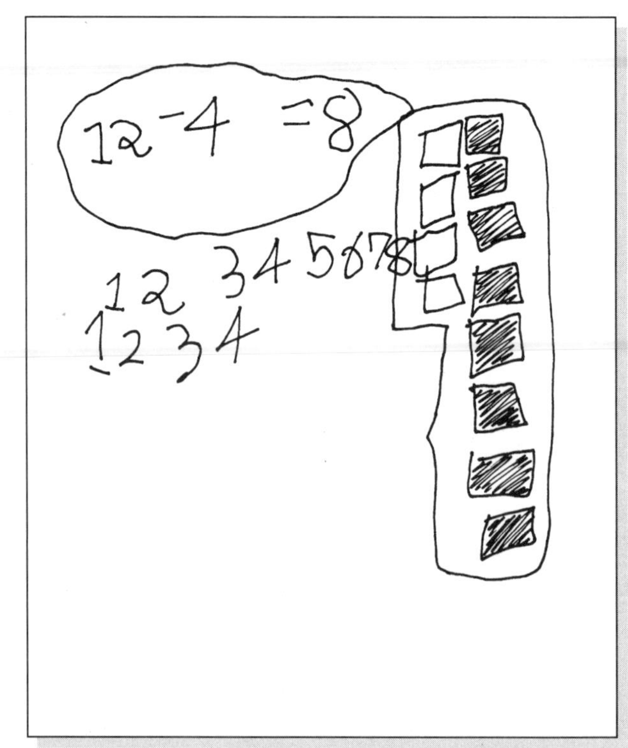

Using Numerical Reasoning

As students learn more about number relationships, they begin to be able to solve problems by taking numbers apart into useful chunks, manipulating those chunks, and then putting them back together.

Luis broke 4 into 2 and 2. He then subtracted each chunk separately: 12 – 2 is 10, then 10 – 2 is 8.

I nae thet
12-2=10 I take
away 2 because
it is sapps to
be 4=8

Tamika explained, "I know 4 and 4 and 4 is 12. Two 4's is 8, and then there's 4 in the pencil box."

4+4+4=12
8

These students are using strategies that involve chunking numbers in different ways, rather than counting by 1's. They are able to visualize the structure of the problem as a whole in order to identify number relationships they know that might help them solve the problem. While it is important to encourage strategies such as these, keep in mind that ability to work with chunks greater than 1 develops gradually over the early elementary years. Many first graders will need to continue counting by 1's for most problems. As they build their understanding of number combinations and number relationships over the next year or two, as well as their ability to visualize the structure of a problem as a whole, they will begin developing more flexible strategies.

Teacher Note

The Relationship Between Addition and Subtraction

It is easy for adults to consider two problem situations as the same, because we would solve them in the same way, while they may actually appear quite different to your students. We may assume that certain situations are addition and others subtraction because we are used to thinking of them that way, but we may find that students solve problems in unexpected ways.

For this reason, as you introduce addition and subtraction problems to students, avoid labeling them as one or the other. A critical skill in solving problems is deciding what operation is needed. Further, many problems can be solved in a variety of ways, and students need to choose operations that make sense to them for each situation. For example, students may solve problems that you think of as subtraction by using addition (in fact, many adults also do this). Consider the following problem:

> 14 children were playing in the park. Then 5 children left. How many children were still in the park?

Most of us learned to interpret this situation as subtraction, and we may naturally assume that students should also see it as subtraction. Students who use direct modeling of the actions to solve the problem will probably count out 14, remove 5, and count how many remain. However, there are many other ways to solve this problem:

- counting down 5 from 14 (13, 12, 11, 10, 9)
- counting up from 5 (6, 7, 8, ...) and keeping track of how many numbers are counted
- using knowledge of number combinations and relationships ("I know 14 take away 4 is 10, so take away 1 more, that's 9.")
- using knowledge of tens and ones ("5 and 10 is 15, and 1 less is 14, so it's 9."

Some of these methods are based on subtraction (moving from 14 down to 5), but others are based on addition (moving *up* from 5 to 14). The method chosen depends on a person's mental model of the situation: Do you see this problem as a taking-away situation to be solved by subtraction, as an adding-on situation, or as a gap between two numbers that might be solved by either addition or subtraction, depending on which is easier in the particular situation? Any of these methods are appropriate for solving this problem. Addition is just as appropriate for solving this problem as subtraction, and, for many students, makes more sense.

Five-in-a-Row and Story Problems

What Happens

Students play a challenging version of Five-in-a-Row, a game introduced in an earlier unit that provides practice with single-digit addition pairs, this time with sums up to 20. They play this game and solve story problems for Choice Time. At the end of Choice Time, students share strategies for some of the story problems they solved. Their work focuses on:

- becoming familiar with single-digit addition pairs
- developing strategies for solving combining and separating story problems
- recording strategies for solving combining and separating story problems, using pictures, numbers, words, and equations

Materials

- Student Sheet 21 (1 per student, homework)
- Student Sheets 22–24 (1 of each per pair and a few extras, plus 1 of each per student, homework)
- Transparency of Five-in-a-Row Board A
- Overhead projector
- Number Cards (1 deck per pair)
- Counters, cubes, or coins (20 per pair)
- Story Problems, Set A (in prepared envelopes)
- Unlined paper
- Paste or glue sticks

Activity

Five-in-a-Row with Three Cards

Students may be familiar with Five-in-a-Row, a bingo-like game played in the unit *Building Number Sense.* Here they play a more challenging version of the game, Five-in-a-Row with Three Cards.

Gather students for a demonstration game, using the overhead and transparency of Five-in-a-Row Board A, and a deck of Number Cards.

Shuffle the Number Cards and turn up the top three cards. Write these three numbers where everyone can see.

I turned up 3, 7, and 1. We can use any *two* of these numbers for the first turn. What's one sum we could make with two of these numbers?... Chanthou added the 3 and 7 and got 10. How did you figure that out? Who has another way to find the sum of 3 and 7?

Record an equation to show the first sum. Then ask students to find other sums of two of the three numbers, and explain how they found them. Record an equation for each sum.

Turn attention to the gameboard and explain that players can place a marker on any square that matches one of their sums. Ask a volunteer to choose a square and place a marker on the board on the overhead.

Turn over three more cards, record them, and again ask students about sums they could make with any two cards. Before you call on someone to place a marker, remind the class that the goal is to mark five squares in a row horizontally, vertically, or diagonally. Encourage students to think about where they could place the marker to help them make five in a row. Be sure they understand that they can cover only one square on each turn.

Continue playing for a few more turns, or until you think everyone understands the game. If the following situations do not arise as you play, explain what students are to do in each:

- If students draw a 0 card, they can use it as one of the addends in any combination. For example, if they turn over 0, 4, and 6, they can make 4 (0 + 4), 6 (0 + 6), or 10 (4 + 6).
- If no move can be made (all possible sums are covered), students turn over three more cards.
- If all the cards in the deck have been used, they reshuffle the deck and turn it facedown again.

Note: As you demonstrate the game, you might model the following method of mixing up the cards in the deck. Keep the cards in three piles as you turn them over. For example: You turn over 9, 1, and 6 and lay them out beside each other, faceup. The next three cards you turn over are 0, 4, and 5. Put the 0 on top of the 9, the 4 on top of the 1, and the 5 on top of the 6. Continue putting each new round of three cards faceup on the three piles. When all the cards are used, stack one pile on top of the other, turn them over, and begin again. No reshuffling is necessary, because the triples used the first time through the deck have already been separated. This takes a long time to explain in words, but it is simple to demonstrate, and it's much easier for students than shuffling.

Activity

Choice Time

Announce Choice Time for the rest of today (Session 3) and most of the next two sessions. Post the two choices.

Halfway through Session 5, you will be calling students together to share strategies for two or three of the story problems. Decide which problems to use for this discussion (for example, problems 1, 3, and 4, or others appropriate for your class). Explain that everyone must solve these two or three problems by the end of Choice Time, though many students will have time to solve the other problems as well. At the start of Session 5, remind students of the problems they must complete.

Since Five-in-a-Row will be fresh in students' minds, you may want to spend more time helping groups of students get started on story problems.

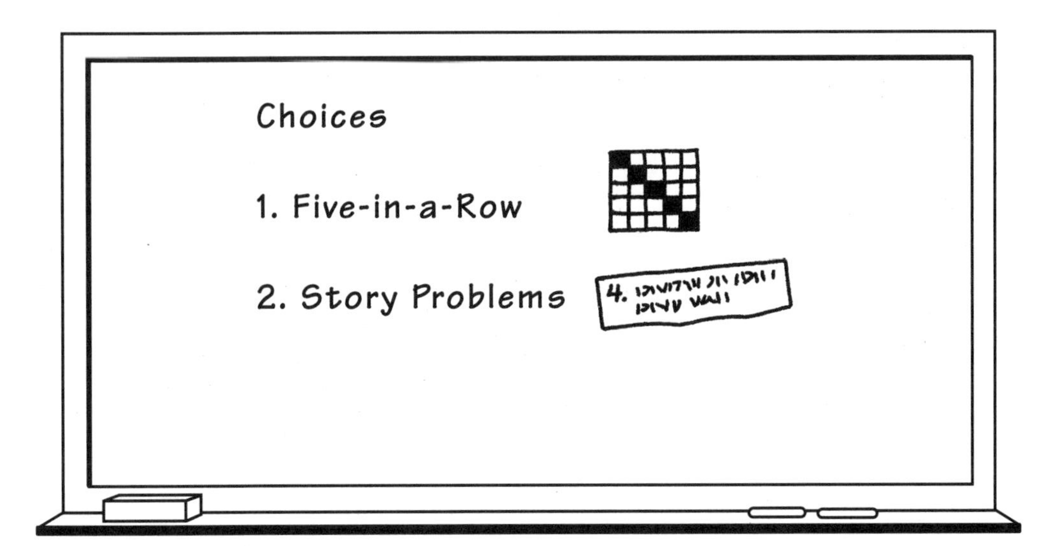

Choice 1: Five-in-a-Row with Three Cards

Materials: Student Sheets 22–24, Five-in-a-Row Boards A–C (1 of each per pair); Number Cards with wild cards removed (1 deck per pair); counters, cubes, or coins to use as markers (about 20 per pair)

Student pairs play cooperatively. Each pair needs a deck of Number Cards, a gameboard, and 20 counters to use as markers. (Players may also use counters to help them find sums.) On each turn, the two players turn over the top three cards in the deck. Together, they choose a sum they can make with any two of the three cards and place a marker on a square with that sum. Any time no move can be made (all possible sums are covered), they turn over three more cards. If they use all the cards in the deck, they mix the cards and turn the deck over to begin again. The game is over when they have covered five squares in a row, either horizontally, vertically, or diagonally.

If students have difficulty keeping counters on the correct squares, they might draw an X to cover a sum, but this will necessitate a new gameboard for each game.

Students may trade boards with other pairs between games, so they play on different boards.

If choosing among different combinations seems too difficult for some students, suggest that they play a few games in which they draw just two cards on each turn. They find the sum of those two cards and place a marker on a square that matches that sum. (This is the simpler version of the game, as presented in the unit *Building Number Sense*.)

Students ready for more challenge might play a version of the game in which they use wild cards. A wild card can stand for any number between 0 and 10. Or, they might play a version in which they turn over five cards on each turn and make sums with any two of the five cards.

Choice 2: Story Problems

Materials: Story Problems, Set A, sorted into seven envelopes; counters (available); unlined paper; paste or glue sticks

Students select one problem at a time from the envelopes of problems you have prepared. They paste it onto the paper they will use. Students work in pairs or individually. They may solve the problems in any way that makes sense to them. When they have found a solution, they record their work so that someone else can understand their solution strategies. Some teachers ask students to work individually at times and in pairs at other times; that way, students gain experience both working alone and collaboratively on recording their work.

You will need to set up procedures for getting help in reading the problems. You might designate certain students as reading helpers, or start off a group of students yourself by reading a problem with them.

❖ **Tip for the Linguistically Diverse Classroom** If you have the support of second-language parents or other translators, this activity will go more smoothly. Otherwise, provide a set of story problems with simple drawings added to help make them comprehensible, or act out each problem with concrete materials.

Also set up procedures for how students can keep track of which problems they complete. They might write Set A at the top of a recording sheet and then write the number of each problem as they complete it. They can add to this sheet in later Choice Times when they work on problems in Sets B–D. Alternatively, you might provide a sheet that lists the story problems in this investigation, and students can check off or circle the problem numbers as they complete them (see example on p. 128).

The **Teacher Note,** About the Story Problems in Investigation 3 (p. 125), describes the types and level of difficulty of the problems in Set A (as well as Sets B–D) and suggests ways to structure student work on them. As needed, guide students in selecting problems at the appropriate level of challenge. If you noticed some students having difficulty with the problems in Sessions 1 and 2, you may want to adjust the numbers in the problems for those students.

Students may continue working on the problems for homework. Those who do not finish all seven Set A problems during Sessions 3–5 may work on them during the next Choice Time in Sessions 6–8.

Observing the Students

As students are working on Choice Time activities, observe and listen to them.

Five-in-a-Row with Three Cards

- How do students find sums? Do they count out each quantity? Do they count on from one of the numbers? Do they use knowledge of number combinations? Can they quickly add 1 or 2 to another number mentally, or do they need to count it out? Do they use number combinations they know to figure out those they don't know?
- How do students determine which combinations to use? Do they seem to choose two of the numbers at random? Do they find all the possible combinations and then choose the one they think will help them get five in a row? Do they decide on a square they would like to cover and then see if they can make the appropriate total with two of the numbers they have drawn?

Story Problems

- Do students understand whether quantities are being combined or separated? Do they know what they need to find in order to solve the problem? Can they keep the situation in mind as they solve the problem?

  Remind students to think through the situation before they solve the problem. How does the problem start? Then what happens? Are there more or less at the end of the story problem than at the beginning? Encourage them to act out the problem with a partner, to build a model of the situation with counters, or to draw pictures of the situation.

- What strategies do students use? Do they count out each quantity from 1? Do they count on or count back from one number? Do they use number combinations they know to figure out those they don't know? Are they beginning to take numbers apart in ways that help them solve the problem more easily? (For example, to add 8 + 5, do they break 5 into 2 and 3?)
- Are students beginning to use strategies that involve tens and ones? ("I know that 13 + 7 is 20 because 3 + 7 is 10, and then 10 more is 20")? Can they quickly add or subtract 10, or do they need to count it out?

  A few students will probably begin using tens in their addition and subtraction strategies. Encourage these students to explain their thinking.

  So, when you were adding 13 and 7, why did you add 3 and 7 first? Then, what was left to add? How far is it from 10 to 20?

  However, do not insist that these students abandon counting by 1's. Most first graders will need to continue counting by 1's as a way of validating their work and a way of approaching challenging problems.

  As you see students who are beginning to do less counting by 1's, you might ask them to repeat a problem with different numbers, about the same size or a little smaller. That way, they can develop their strategies while working with familiar numbers that they are comfortable breaking apart and putting together in different ways.

- How clearly can students show their strategies? Are students using numbers or equations as part of their recording?
- Are the numbers in the problems of average difficulty for most students? Do some students need smaller or larger numbers? As you circulate, you can modify the numbers where appropriate.
- Do students have ways to prove that their solutions are correct?

If there is a particular problem that is difficult for several students, you might pull them together to work with you as a small group. Encourage students to think through the whole problem before they try to solve it. They can use pictures or counters to model the problem for themselves.

Students who are ready for more challenge might find and record more than one way to solve each problem.

Set A, problem 1: Ken found 12 white shells at the beach. He found 6 brown shells. How many shells did he find?

12 + 6 = 18 the 12 take away the 2 from you have 10 and 6 + 2 = 8 and 10 + 8 = 18

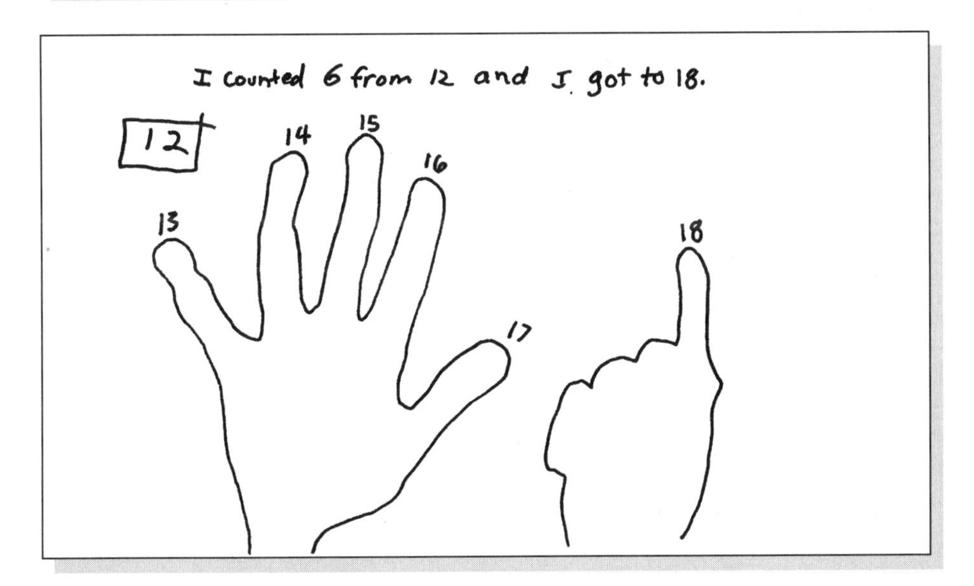

Set A, problem 3: The class needs boxes. Tara brought 7 boxes. Ken brought 4 boxes. Then Tara brought 3 more boxes. How many boxes do they have now?

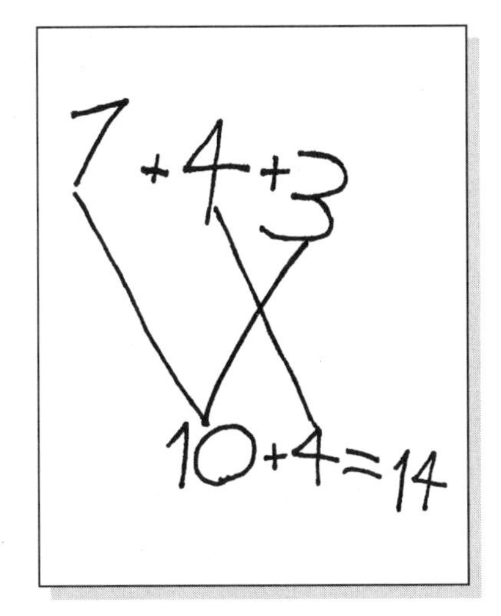

7 boxes 4 boxes

1 2 3 4 5 6 7 8 9 10 11

3 more boxes

12 13 14

= 14

Activity

Sharing Story Problem Strategies

About halfway through Session 5, call students together to share strategies for the problems everyone was asked to complete. Gather several strategies for each problem, recording each method on the board or chart paper. Encourage students to think about each others' ideas by occasionally asking for volunteers to restate someone else's strategy. For example, after Max explains his strategy, you might ask:

Who can say in your own words how Max did this problem?

As students share their work, discourage any apparent reliance on individual words in the problem that might seem to signal a particular operation. The **Teacher Note**, "Key Words": A Misleading Strategy (p. 165), explains the drawbacks of this approach.

Set A, problem 4: Two goats had 15 carrots. They ate 8 of them. How many carrots were left?

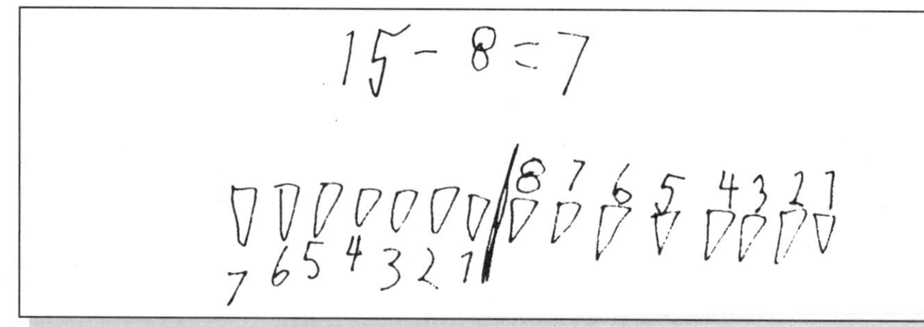

Sessions 3, 4, and 5 Follow-Up

Homework

Story Problems Students may continue to work on story problems from Set A for homework. Ask them to paste or glue each problem to a sheet of paper before they take it home.

Five-in-a-Row with Three Cards Students teach someone at home to play Five-in-a-Row with Three Cards. They will need the directions on Student Sheet 21, a deck of Number Cards, Five-in-a-Row Boards A–C, and counters (such as buttons or pennies) to use as game markers.

About the Story Problems in Investigation 3

Teacher Note

Three sets of story problems are used during Choice Times in this investigation, with an optional fourth set provided for students who are ready for more challenge.

Problem Types Most of the story problems in this investigation, like those in the earlier grade 1 unit *Building Number Sense,* involve combining and separating with an unknown outcome. (To review these problem types, see the **Teacher Note,** Types of Story Problems, p. 108.) Making sense of and developing strategies for solving these types of problems is an important part of the number work that first graders do in the *Investigations* curriculum. Also included are some problems with unknown change, which are often more challenging for first graders. While first graders need to become familiar with different problem structures, it is not necessary for everyone to gain proficiency with problems involving unknown change during first grade.

Number Range The problems in Sets A–D involve a range of numbers in order to help students build and extend their addition and subtraction strategies.

Some problems use small, familiar numbers, such as 8, 12, and 15. While many first graders will initially approach these problems with direct modeling or counting strategies, some students, with experience, gradually begin using more of what they know about number combinations to find their solutions. For example, you might hear them reason that "8 and 9 is 17 because you can take 1 from the 8, add it to the 9 and get 10, and then 7 more is 17." Students who are starting to do less counting by 1's will benefit from doing more problems with familiar numbers.

Some problems have numbers or totals in the 20's and 30's. A few students approach these problems with strategies that involve number combinations or chunking numbers in tens and ones. ("I know 16 and 14 is 30 because 6 and 4 is 10, 10 and 10 is 20, and 10 more is 30.") But, most first graders approach these problems with strategies that involve counting by 1's. As they draw or count out the quantities in the problems and as they share strategies with their classmates, they develop a better understanding of these larger numbers and relationships among them. In the next year or so, as their understanding grows, they too will probably begin using some numerical reasoning to solve these problems.

You may decide to adjust the numbers in the problems for some or all of your students. If you have a few students who are comfortable reasoning about number combinations, challenge them with larger numbers. Students who continue to use counting strategies and have difficulty keeping track of the counts, and those who are just beginning to use strategies that involve number combinations, can continue to work with the numbers the same size or smaller.

A summary of the types of problems in each set follows. If you think some students are having difficulty, you might suggest they begin with particular problems. Otherwise, it is not necessary for them to do the problems in sequence. Students should not be told what types of problems these are, as they need to think of the situation the problem is describing and then determine themselves how best to solve it.

Continued on next page

Teacher Note

continued

Story Problems, Set A

These problems are first used in Sessions 3–5.

Problems 1–2: Combining with an unknown outcome. Students find the total of two amounts.

1. Ken found 12 white shells at the beach. He found 6 brown shells. How many shells did he find?
2. Tara had 13 erasers. Then her friends gave her 7 erasers. How many erasers did Tara have?

Problem 3: Combining three amounts with an unknown outcome. Students find the total of three different amounts.

3. The class needs boxes. Tara brought 7 boxes. Ken brought 4 boxes. Then Tara brought 3 more boxes. How many boxes do they have now?

Problems 4–6: Separating with an unknown outcome. Students find how many are left when one amount is removed from another.

4. Two goats had 15 carrots. They ate 8 of them. How many carrots were left?
5. Ken had 16 pennies. He lost 10 pennies. How many did he have left?
6. Tara had 17 stickers. She gave away 9 of them. How many stickers did she have left?

Problem 7: Combining several equal amounts. Students solved problems of this type in Investigation 2.

7. I see 4 children and 2 dogs with muddy feet. How many muddy feet do I see?

Story Problems, Set B

These problems are first used in Sessions 6–8.

Problems 1–2: Combining with an unknown outcome.

1. Shani had 15 beads on a string. She added 10 more beads to the string. Now how many beads does she have?
2. Shani washed 19 paint brushes. Alex washed 8 brushes. How many brushes in all did they wash?

Problem 3: Combining and separating with an unknown outcome. Several amounts are given. Students interpret the problem situation to find which amounts are to be combined and which are to be separated, and they find the result.

3. Alex had 15 pennies in one pocket and 6 pennies in his other pocket. He spent 5 pennies on a sticker. How many pennies did he have left?

Problems 4–6: Separating with an unknown outcome.

4. Mr. Wing had 14 pumpkins. He sold 11 of them. How many pumpkins were left?
5. Shani had 25 balloons. She gave away 7 of them. How many balloons did she have left?
6. Alex made 22 tacos for a party. His friends ate 12 tacos. How many tacos were left?

Problem 7: Combining several equal amounts.

7. There are 5 children at one table. Each child has 5 pencils. How many pencils do they have in all?

Story Problems, Set C

These problems are first used in Sessions 10–12. They include some combining with unknown change. It is important not to distribute these problems until after the whole-class introduction to combining with unknown change in Session 9.

Problems 1–3: Combining with unknown change. A total and one amount are given, and the second amount must be found.

1. Tara had 5 crayons. Her father gave her some more. Now she has 9 crayons. How many crayons did her father give her?
2. Ken had 8 marbles. Then he found some more. Now he has 12 marbles. How many marbles did he find?
3. 10 people were on a bus. Some more people got on. Now there are 15 people on the bus. How many people got on the bus?

Problem 4: Combining with an unknown outcome.

4. Tara and Ken collected 16 cans. The next day they collected 14 more cans. How many cans did they have?

Problem 5: Combining and separating with an unknown outcome.

5. Tara picked 12 apples from one tree. She picked 13 apples from another tree. She gave 10 apples to Ken. Then how many apples did Tara have?

Problem 6: Separating with an unknown outcome.

6. 30 children were playing ball. Then 11 of them went home. How many children were still playing ball?

Problem 7: Combining several equal amounts.

7. Tara has 2 boxes of 10 markers and 2 boxes of 5 markers. How many markers does Tara have in all?

Story Problems, Set D (Challenges)

Reserve these problems for students who complete all of Sets A, B, and C and are ready for extra challenge.

Problems 1–2: Combining with unknown change.

1. There were 13 black birds in a tree. Some red birds flew into the tree. Now there are 20 birds in the tree. How many red birds are in the tree?
2. Alex's train has 18 cars. He added some more cars to it. Now the train has 25 cars. How many cars did Alex add?

Problems 3–4: Separating with unknown change. A total and one amount are given, and the second amount must be found. Note that there is no whole-class introduction to this challenging problem structure.

3. Shani gave her rabbits 13 carrots. The rabbits ate some. Now there are 9 carrots. How many carrots did her rabbits eat?
4. Shani had 16 pennies in her pocket. She had a hole in her pocket. Some pennies fell out. Now she has 7 pennies. How many pennies fell out?

Problem 5: Combining with an unknown outcome.

5. The children in Center School have 23 cats and 12 dogs. How many pets do they have?

Problem 6: Separating three amounts with an unknown outcome.

6. Alex baked 30 cookies. His father ate 6 cookies. His brother ate 5. His sister ate 4. How many cookies were left?

Problem 7: Combining and separating with an unknown outcome.

7. Alex had 42 cents. He lost 15 cents. Shani has 10 cents. Do they have enough money to buy a sticker that costs 30 cents? How do you know?

Continued on next page

Teacher Note

continued

Guiding Student Work For each Choice Time, specify two or three problems for all students to attempt; these problems will be the focus of a whole-class discussion at the end of that Choice Time. Students may choose freely among the other problems (including any not completed from previous Choice Times). However, at times you may want to assign particular problems to students who will not have time to complete all the problems, and who could use more work with certain problem types or with numbers of a certain size.

To help students who are having difficulty, you might call together small groups to work with you on a problem during Choice Time; you might adjust the numbers in some problems; or you might decide to create some problems of your own. Remember that while it's fine to alter the numbers or the context of the problems, even a slight change in the problem structure can mean a great change in the level of difficulty.

To help students keep track of which problems they have done, you might provide a simple form. If you also provide story problems you have written, you can include these on the same sheet.

I did these story problems:

Set A
1 2 3 4 5 6 7

Set B
1 2 3 4 5 6 7

Set C
1 2 3 4 5 6 7

Set D
1 2 3 4 5 6 7

Sessions 6, 7, and 8

Tens Go Fish

What Happens

In the game Tens Go Fish, students make combinations of ten with two addends. This game is added to the previous choices (Five-in-a-Row and story problems) as Choice Time continues for three more sessions. A whole-group round of Quick Images starts Session 7, and during the last half of Session 8, students share strategies for some of the story problems they solved. Their work focuses on:

- becoming familiar with combinations of 10
- reasoning about combinations of 10
- becoming familiar with single-digit addition pairs
- developing strategies for solving combining and separating story problems
- recording strategies for solving combining and separating story problems, using pictures, numbers, words, and equations

Materials

- Number Cards (1 deck per pair)
- Prepared card holders (optional)
- Story Problems, Set B (in prepared envelopes), and any remaining in Set A
- Unlined paper
- Paste or glue sticks
- Cubes or other counters (available)
- Transparencies of Quick Image Squares or Dot Addition Cards
- Overhead projector
- Student Sheet 25 (1 per student, homework)

Activity

Introducing Tens Go Fish

If your students are familiar with the card game Go Fish, they will need just a brief introduction to Tens Go Fish.

First explain about "making 10" with pairs of Number Cards. Draw five Number Cards in a row on the board or on chart paper. Include one pair that makes 10. For example, you might select the cards 4, 1, 5, 7, and 9.

I'm going to show you a game called Tens Go Fish. The object is to find pairs of cards that add up to 10. Each player gets five cards to start. Let's say these cards are the cards in my hand: 4, 1, 5, 7, and 9. Can I make 10 with two of these cards?... OK, I could make 10 with the 1 and the 9. That's my first pair.

Redraw the 1 and 9 cards, as a pair, to one side.

If I look at my hand when the game starts, and I have a pair that make 10, I can take them out and then draw two more cards.

Replace the cards you have put down with two more cards; this time be sure that no pairs of cards in your hand make 10. For example, if you have 4, 5, and 7, you might add another 4 and a 2.

Let's say I drew a 4 and a 2, so now these are my cards: 4, 2, 4, 5, and 7. Do any two of these cards make 10?

When it's my turn, I can ask the other player for a certain card that I need to make a total of 10. For example, suppose I wanted to make 10 using the 2 in my hand. What card would I need to add to the 2 to make 10?

So, if I was playing with Claire, I might ask, "Claire, do you have an 8?" If Claire has an 8, she gives it to me. I put the 8 and the 2 down as a pair, and draw the top card from the deck. If Claire does not have an 8, she says "Go Fish." I take a card from the top of the deck.

Each time I draw a new card, I check to see if I can make 10 with that card and one already in my hand. If I can, I put the pair aside and draw a new card. If I can't, my turn is over.

Start a demonstration game with a student volunteer. Explain that for this game, you will be showing students your cards so that they can learn how to play. When students play in pairs, they will not show their cards to their partners. As you play, involve students in your turn.

I have a 5, 7, 2, 1, and 4. Can I use two of these cards to make 10? No one sees a way? OK, so what could I do next?

You might decide to play an entire demonstration game, or if you think most students understand how to play, just play for a few turns. In this case, explain that the game continues with each player trying to make combinations of 10. The game is over when there are no more cards.

If you never have a 10 and a 0 card in one hand during the demonstration game, find these cards and be sure students recognize that they can make a pair with 10 and 0.

As you collect pairs that make 10, put each one in a separate pile. Explain that this is so the cards don't get mixed up, because at the end of the game, players turn over their pairs and list all the combinations of 10 they made, using addition notation. Model this for your students.

Note: If your Number Cards are duplicated on paper, students should use card holders for this game to prevent numbers from showing through.

Activity

Teacher Checkpoint

Choice Time

For the rest of Session 6 and much of next two sessions, students work on Choice Time. Post the three choices.

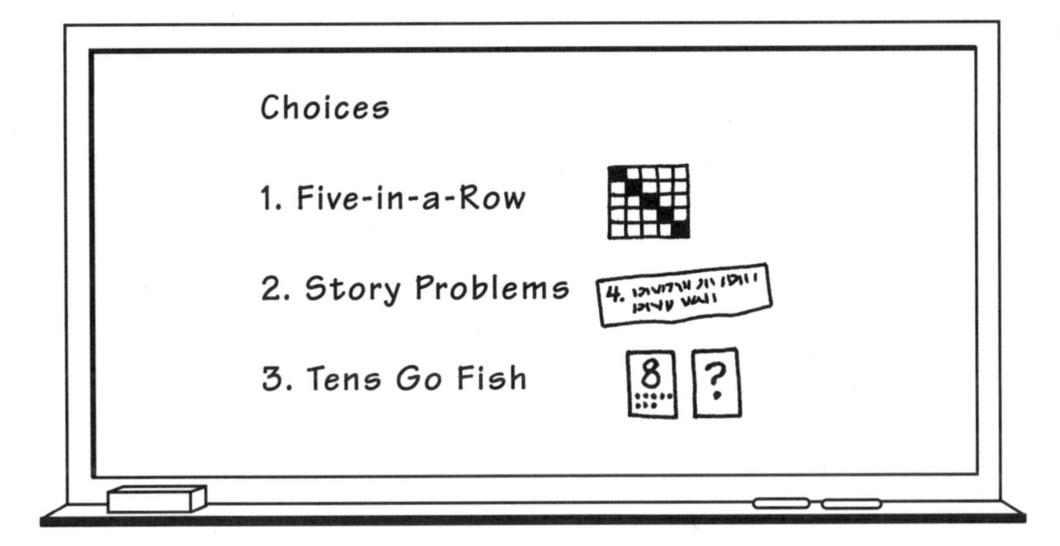

Students need to complete their work on Five-in-a-Row during this Choice Time.

The Story Problems activity is a Teacher Checkpoint, so be sure to observe all students at work on this second choice, to get a sense of the strategies they are using.

Start Session 7 with a whole-class round of Quick Images. Halfway through Session 8, call students together to share strategies for two or three story problems; as that session begins, remind them of the problems they are to complete.

For a description of Choice 1: Five-in-a-Row with Three Cards, see page 120.

Choice 2: Story Problems

Materials: Story Problems, Set B, sorted into envelopes, and any remaining problems from Set A; counters; unlined paper; paste or glue sticks

Students continue solving and recording solutions to story problems. Choose two or three of the problems in Set B, such as 2, 3, and 5, for all students to attempt. Halfway through Session 8, call students together to share their strategies for these problems.

Because the Set B problems involve numbers in the teens and 20's, a few students may begin using strategies involving tens and ones, especially for combining problems. At the same time, some students who have begun using stategies involving number relationships for problems with smaller numbers may return to counting or direct modeling strategies when faced with slightly larger numbers. As you observe students at work, make note of any who are using strategies that involve tens and ones.

Most students will have time to work on several problems. You might guide them in selecting problems at the appropriate level of challenge, or let them make their own choices. Students may continue working on the problems for homework. Students who do not finish all of Sets A and B during Sessions 6–8 might continue working on them during the next Choice Time in Sessions 10–12.

If you noticed some students having difficulty with the problems in Set A, you may want to adjust the numbers in Set B.

For more challenge, students can find more than one way to solve the problems.

Choice 3: Tens Go Fish

Materials: Number Cards with wild cards removed (1 deck per pair); card holders (optional); unlined paper; counters (available)

Students play in pairs or threes. Each player is dealt five cards. (Use card holders, made as described on p. 101, if the numbers show through to the back of your cards.) Players take turns asking each other for cards that will make 10 with a card already in their hand. They place any pairs made on the table and draw a new card from the deck at the end of each turn. If a card drawn from the deck makes a pair with a card in the hand, the player puts that pair down and draws again.

If a player uses up all his or her cards and there are still cards left in the deck, that player draws two cards. The game is over when there are no more cards in the deck. At the end of the game, players list the combinations of 10 they made.

For more challenge, students can play the game in groups of three or four; with more players, it is more difficult to remember the cards other players have asked for.

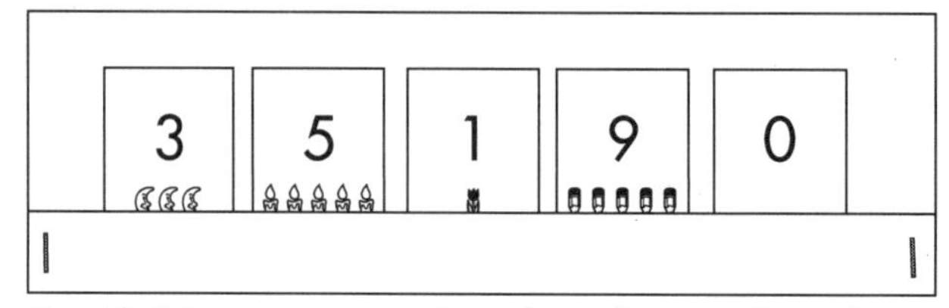

Card holders prevent numbers from showing through paper cards.

Observing the Students

For guidelines on observing students' work on Choice 1: Five-in-a-Row with Three Cards, see p. 121.

Story Problems

Observe during this checkpoint activity to get a sense of how students are solving combining and separating problems. Make notes on each student, jotting down which problems they are doing successfully, which cause them difficulty, and what strategies they are using. If possible, observe each student working on a combining problem and a separating problem.

At the end of Session 8, collect students' work on story problems. Reviewing their written work in conjunction with your observation notes will tell you which students are counting all, which are counting on, which are building on number relationships they know, and which are breaking up numbers in flexible ways.

Set B, problem 5: Shani had 25 balloons. She gave away 7 of them. How many balloons did she have left?

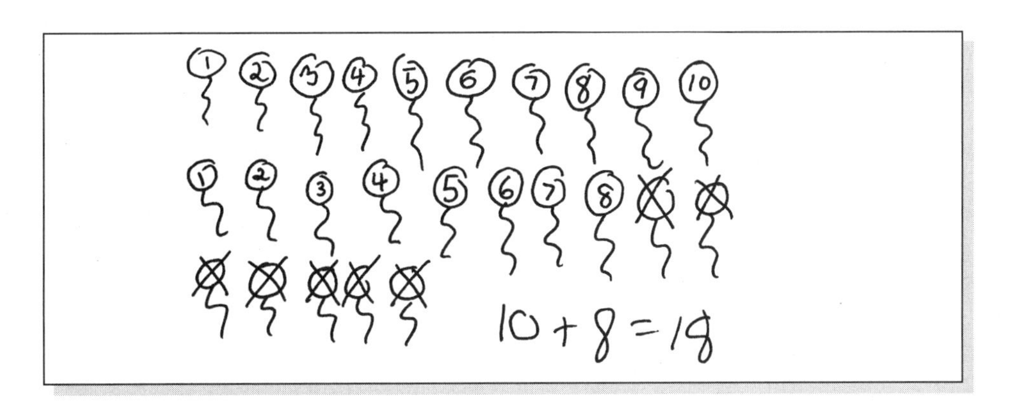

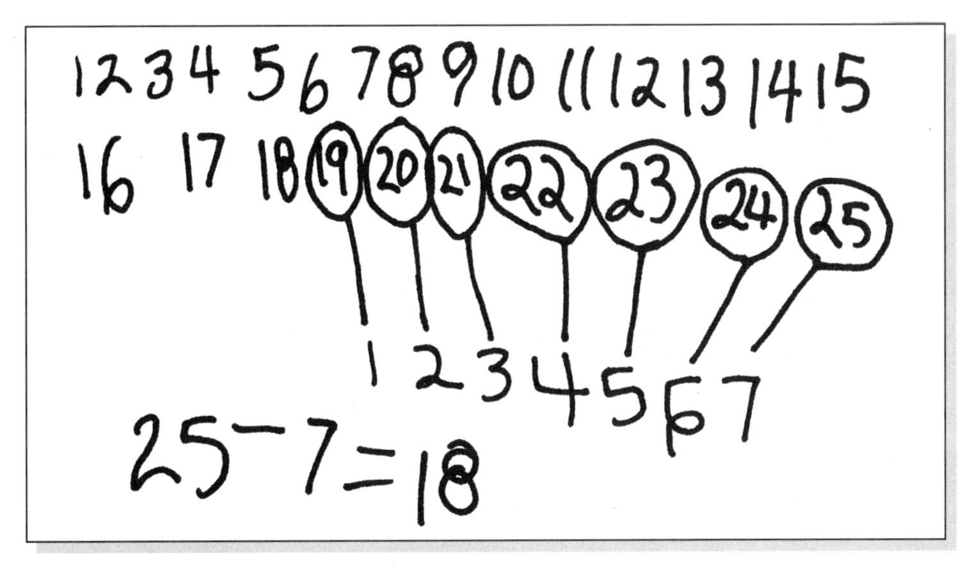

For example, consider Set B, problem 2: "Shani washed 19 paint brushes. Alex washed 8 brushes. How many brushes in all did they wash?" Look for the following in student work:

- direct modeling strategies (such as counting out or drawing pictures of a group of 19 objects and a group of 8 objects, then counting them all from 1)
- counting strategies (such as counting up 8 from 19: "20, 21, 22, ... 26, 27")
- strategies involving number combinations or tens and ones, such as these:

  "I took 1 from the 8 and put it with the 19 to get 20. Then, there's 7 left. I know 20 and 7 is 27."

  "19 is 10 plus 9. I know 9 and 8 is 17. Then, 10 more is 27."

I gave 1 to the 19 and there was 7 more left from the 8. And 1+19=20
20+7=27

I used my head.

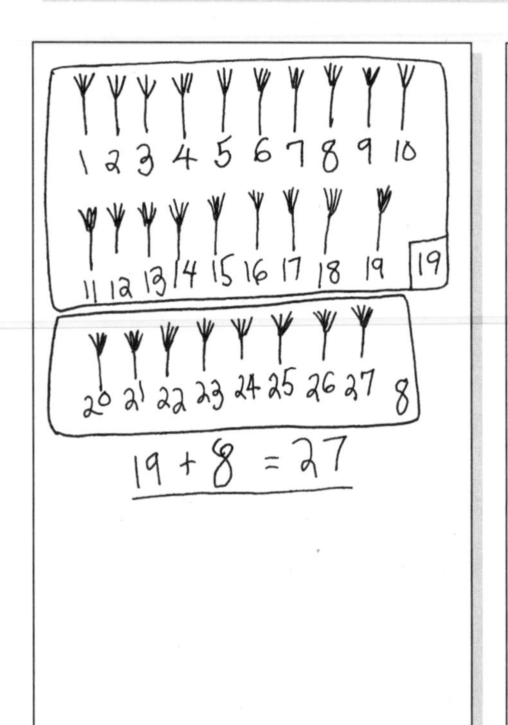

19 8

10 9 8

10 17

10+10=20

20+7=27

Set B, problem 7: There are 5 children at one table. Each child has 5 pencils. How many pencils do they have in all?

I couinted byfivs.

5 10 15 20 25 pencils

Through this checkpoint, think about the class as a whole as well as what individual students may need as they work on the rest of the story problems in this unit. For example, some students who are using direct modeling or counting strategies for the Set B problems might use strategies involving number combinations and relationships if the problems had smaller, more familiar numbers. If most students use strategies involving counting by 1's to solve these problems, consider including a couple of problems with smaller numbers (such as 16 – 9 or 7 + 8) in the next Choice Time.

Note: As students work on the story problems in this investigation, algorithms (procedures) for "carrying" and "borrowing" may come up if students have learned them elsewhere. They must be able to justify and explain these procedures, just as they explain their other strategies. Work with any students who use these approaches to see if they know how the steps in their algorithms relate to what is going on in the problem. If they can recite the steps but cannot relate them to what they are trying to find out, or if they frequently lose track of the procedures, suggest that they find another approach. Throughout the *Investigations* curriculum, it is important that students develop computational strategies that are firmly grounded in their own understanding of numbers and operations.

Tens Go Fish

- How do students decide which card to ask for? Do they use knowledge of combinations of 10? Do they use counting strategies to find a number that goes with a card in their hand to make 10? Do they seem to ask for cards at random?
- Are students able to keep track of the cards other players have asked for? Do they use this to reason about what cards the other player has?

  Some pairs might benefit from playing cooperatively. After a player chooses one card to use to make 10, both players figure out together which other card is needed to finish the pair. If the other player does not have this card, the pair can look for another way to make 10, using one card from each hand.

Activity

Quick Images

At the start of Session 7, play two or three rounds of Quick Images with the class. (See p. 7 to review the steps of Quick Images.) You might use images from Quick Image Squares, or combinations of Dot Addition Card transparencies:

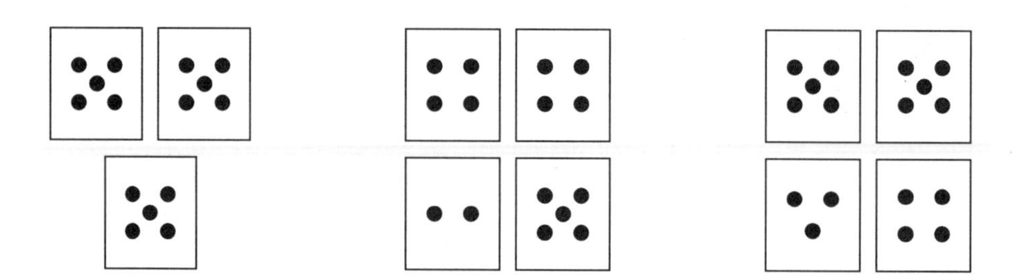

Activity

Sharing Story Problem Strategies

About halfway through Session 8, call students together to share strategies for the problems in Set B you asked everyone to complete. Gather several strategies for each problem. If you saw any students using strategies involving tens and ones, encourage them to share their work, but also call on students who have used other strategies, such as direct modeling or counting on or back. Record on the board or chart paper each strategy that students suggest, basing your recording on the student's own method.

Sessions 6, 7, and 8 Follow-Up

Story Problems Students continue to work on story problems from Set A and Set B that they have not had a chance to finish during Choice Time. Each problem should be glued to a sheet of paper before it goes home. Students should bring this work in to keep with the other story problems they have solved.

Tens Go Fish Students teach someone at home to play Tens Go Fish. They will need directions (Student Sheet 25, Tens Go Fish), and a deck of Number Cards.

Session 9

Combining with Unknown Change

Materials

- Story problem that involves combining with unknown change, written on chart paper
- Interlocking cubes (class set)
- Unlined paper
- Student Sheet 26 (1 per student, homework)

What Happens

Students work with another type of story problem, "combining with unknown change" (a total and one of the amounts are given, and the second amount must be found). Again, they record and share their solution strategies. Then, for homework, they write their own story problems to match a given addition expression. Their work focuses on:

- visualizing what happens in combining with unknown change situations
- developing strategies for solving story problems that involve combining with unknown change
- recording strategies for solving story problems, using pictures, numbers, words, and equations
- creating story problems to match addition expressions

Activity

Combining with Unknown Change

This activity introduces problems that involve combining with unknown change, sometimes referred to as "missing addend" problems. Neither term (*missing addend* or *unknown change*) should be used with students, as they need to work out for themselves what information is needed to solve these problems.

Tell students a story problem (based on the one you have recorded on chart paper) and ask them to try to visualize what happens. If you decide to make up your own story, keep the basic structure and the numbers of the story provided here. Many students find combining with unknown change to be difficult, so the story should be one that students can easily model or act out. It is also essential to use small numbers, with a difference of no more than 2 or 3, so that students can focus on making sense of the problem situation.

❖ **Tip for the Linguistically Diverse Classroom** Have blocks or cubes available as a visual reference; students may want to use them to model the problem.

As you tell the story, some students may want to close their eyes as they try to imagine what's happening.

The other day, I was visiting my neighbor. Her little girl was building a tower out of blocks. She built a tower with 8 blocks. Then, she put some more blocks on the tower. She ended up with 11 blocks in the tower.

Ask for several volunteers to retell the story in their own words. When you're satisfied that students can imagine the situation, ask them what information is missing in the problem, or what they need to figure out. Discourage students from giving an answer at this point.

Some students who remember the story correctly may nonetheless say that what they need to figure out is "how many blocks in all?" They recognize that the problem involves combining two amounts somehow, and in their experience, in order to solve such problems, you need to find a total. Encourage these students to think carefully about the sequence of actions in the problem.

Can you tell me what happened in the story? Do we know how many blocks she started out with? Then what happened? OK, she got some more. After she got some more, how many did she have? So, what do you think we need to find? Do you think the answer will be more than 11? Why not?

Show students the problem as you have recorded it on chart paper. Make available unlined paper and interlocking cubes. Students now solve this problem and record their thinking in a way that helps someone else understand how they found their solutions.

> A girl built a tower with 8 blocks. She put some more blocks on the tower. Then there were 11 blocks in the tower. How many blocks did she put on the tower?

Many students will probably count out or draw eight things, then keep track of how many more they need to count out or draw until they have 11. Some students may count up from 8 or back from 11, and a few students may use knowledge of number combinations: "I know she added 3 blocks, because 8 and 2 is 10, and 1 more is 3," or "It's 3 because 8 + 3 is 11."

Some students may need lots of help to make sense of the problem and keep the problem situation in mind as they are solving it. In the **Dialogue Box,** How Many in All, or How Many Were Added? (p. 142), a teacher helps one student having difficulty with this problem.

Students who finish early can show a second way to find the answer. If you think some students are ready for more challenge, you might pose a problem involving combining with unknown change that uses larger numbers.

> So, there were 11 blocks on the tower. The girl put more on, and now there are 17. How many did she put on?

When students have finished, call the class together to share their strategies and how they recorded them. Have counters and cubes available for demonstration. Record each method suggested on the board or on chart paper.

Repeat the activity with another problem involving combining with unknown change. If you used a story about towers last time, tell a story about something else this time, to give students experience thinking about combining with unknown change in different contexts. If many students found the first problem difficult to make sense of, present a problem with slightly smaller numbers that are only 1 or 2 apart (say, an initial amount of 6 and a total of 8). Otherwise, present a problem at about the same level of difficulty. For example:

My neighbor's children decided to play with marbles. Her little girl had 6 marbles, and her little boy had some, too. They had 10 marbles together. How many did the boy have?

You might record this problem on chart paper for students to refer to as they work.

Activity

Writing a Story Problem

To prepare students for the homework, write an addition expression, such as 5 + 7, on the board or chart paper. Ask students to suggest a story problem for it.

I've written 5 + 7. Who can think of a story problem for 5 + 7? What happens in the problem? What could the 5 be? the 7?

Gather several different ideas. If students are stumped, give an example yourself to get them started. See how many different problem contexts the class can suggest.

Session 9 Follow-Up

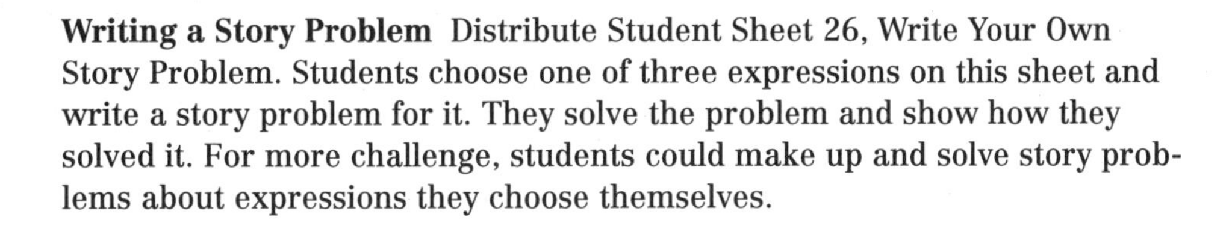

Writing a Story Problem Distribute Student Sheet 26, Write Your Own Story Problem. Students choose one of three expressions on this sheet and write a story problem for it. They solve the problem and show how they solved it. For more challenge, students could make up and solve story problems about expressions they choose themselves.

Name Leah Date April 20

Student Sheet 26

Write Your Own Story Problem

Choose one of the following.
Write a story problem for it.

8 + 7 4 + 5 + 6 20 + 10

Ms. Scott has 8 books on Scotland. Then she has 7 more books On Scotland. How many books on Scotland does she have?

Show how you solved your story problem.
Use words, pictures, or numbers.

I took two from 7 and gave the two to the 8 and there was 5 left from the 7 and the 8 was a 10. 5+10=15

 219 Investigation 3 • Session 9 *Number Games and Story Problems*

Combining and Separating Stories Read aloud the book *Splash* by Ann Jonas (Greenwillow Books, 1985). On each two facing pages, some people or animals fall in or jump out of a pond. Readers then find the total number in the pond. After this reading, ask students to create their own stories that illustrate combining and separating. On each page they might show how many of a set of objects were gained or lost, and the total so far. For example, they might write about pennies being found or lost or spent, or baby birds hopping in and out of a nest.

How Many in All, or How Many Were Added?

The teacher is circulating to observe students working on the problem introduced in class for combining with unknown change.

> A girl built a tower with 8 blocks. She put some more blocks on the tower. Then there were 11 blocks in the tower. How many blocks did she put on the tower?

Eva has recorded the numbers 12, 13, 14, 15, 16, 17, 18, 19 on her paper, and has circled the 19. The teacher asks her about her work.

Can you tell me about how you solved the problem?

Eva: I counted. I counted up 8 numbers.

What were you trying to find?

Eva: How many blocks in all.

OK, show me how what you did fits the problem. *[The teacher reads]* **"A girl built a tower with 8 blocks." So, how did you show that?**

Eva: I did 8 numbers *[points to the numbers she has recorded].*

What happened in the problem after that?

Eva: *[reading]* "She put more blocks on. Then there were 11 blocks in the tower." So I did 11... Oh! She has 11 *in all.*

So, what do we need to figure out?

Eva: How many in all. The answer is 11!

Well, there are 11 in all. She started with 8. *Then* she got some more. She ended with 11.... OK, let's look at the question at the end of the problem.

Eva: *[reading]* **"How many blocks did she put on the tower?"**

[The teacher reaches for a box of cubes and is about to ask Eva to use cubes to model the problem, when the girl announces another idea.]

Eva: She *put on* less than 11! Because she had 11 in all.

So, how could we figure it out?

Eva: It's... 9, 10, 11 *[keeping track on her fingers]...* so 3. She put on 3 blocks.

So, now let's go back to the problem and see if this makes sense. She started with 8, and then how many did she take?

Eva: She took 3, and she had 11 in all, because 8... 9, 10, 11 *[holds up three fingers].*

Addition and Subtraction

What Happens

In the unit's final Choice Time, students work on Total of 20, (a variation of a game from Investigation 1), Tens Go Fish, and story problems. During the last half of Session 12, students share strategies for solving story problems. Their work focuses on:

- developing strategies for solving a variety of combining and separating story problems
- recording strategies for solving combining and separating story problems, using pictures, numbers, words, and equations
- becoming familiar with combinations of 10 and 20
- reasoning about combinations of 10

Note: Find some time either at the start of this session or later in the day to acknowledge students' homework on Write Your Own Story Problem. For example, ask a few students to share their problems during writing time.

Name Andre Date April 20

Student Sheet 26

Write Your Own Story Problem

Choose one of the following.
Write a story problem for it.

8 + 7   4 + 5 + 6   (20 + 10)

Andre has 10 vanilla cookies and Demian has 20 chocholate cookies. How many cookie are there in all?

Show how you solved your story problem.
Use words, pictures, or numbers.

20 + 10
I know that 2 + 1 = 3 and 0 + 0 = 0 so I know that its 30.

 219 *Investigation 3 • Session 9* *Number Games and Story Problems*

Materials

- Number Cards (1 deck per pair)
- Story Problems, Sets C–D (in prepared envelopes), and any remaining in Sets A and B
- Unlined paper
- Paste or glue sticks
- Cubes or other counters (available)

Activity

Choice Time

Most of Sessions 10–12 will be spent in Choice Time. Post the three choices. The new choice, Total of 20, is a variation of the game Total of 10 that students played in Investigation 1. Take a few minutes to review the game and to introduce this more challenging version. See the directions on Student Sheet 7, p. 183. Total of 20 is variation C.

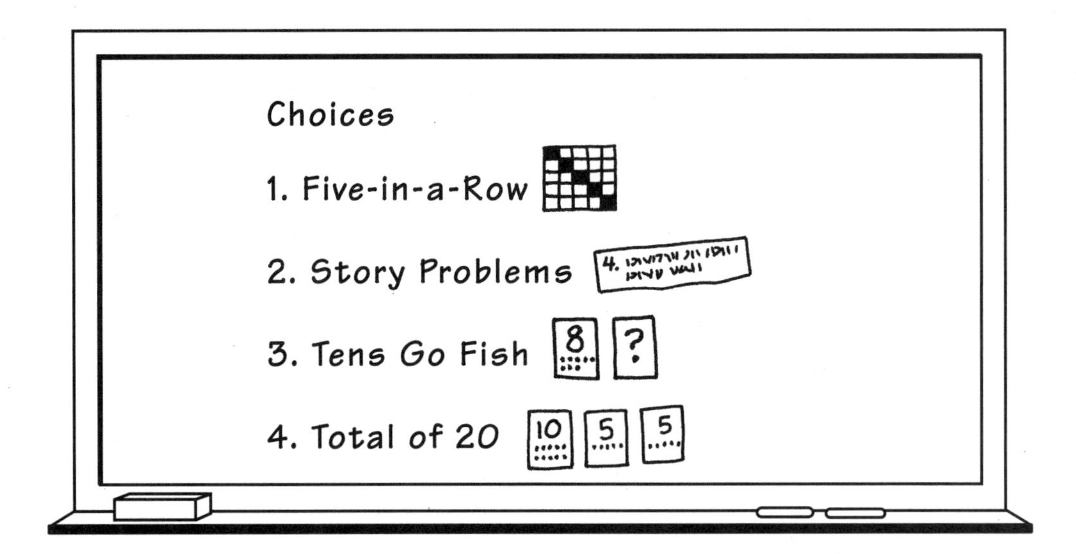

Choice Time ends about halfway through Session 12, when you call students together to share solutions to two or three of the story problems.

To review Choice 3: Tens Go Fish, see page 132.

Choice 2: Story Problems

Materials: Story Problems, Set C, sorted into envelopes, and any remaining problems made from Sets A and B; counters; unlined paper; paste or glue sticks. (Have Story Problems, Set D, available for any students ready for more challenge.)

Students continue solving and recording solutions to story problems. Choose two or three problems in Set C for all students to attempt, including one for combining with unknown change (such as problem 2) and one or two other problems (maybe problems 4 and 5). While all students must solve these two or three problems during this Choice Time, most will have time to complete other problems as well. At the start of Session 12, remind students of the problems they are to complete. Midway through Session 12, the class meets to share strategies for these problems.

If you noticed some students having difficulty with the problems in Session 9, you may want to adjust the numbers in Set C problems 1, 2, and 3.

Some students may not have time to complete all the story problems in this investigation. Consider suggesting that particular students work on certain problems for more practice with problem types they are having difficulty with. Students may continue their work on these problems for homework.

If some students finish all the problems in Sets A–C and they seem ready for more challenge, offer them Story Problems, Set D (Challenges).

Choice 4: Total of 20

Materials: Number Cards (1 deck per pair); unlined paper; counters (available)

Students play in pairs. If necessary, three students can play together or a student can play alone. Players start by laying out 20 cards, faceup, in four rows of five.

Total of 20 is played much like Totoal of 10, except that players take turns looking for combinations of numbers that total 20. The other difference is that in this version, each time players set aside a combination of 20, they replace those cards in the layout with more from the deck. The game is over when no more combinations of 20 can be made. Players then list the combinations of 20 that they made.

Students ready for more challenge can play with wild cards, which can be used as any number.

Observing the Students

For guidelines on observing students' work on Choice 3: Tens Go Fish, see p. 136.

Story Problems

Pay particular attention to how students are solving the problems that invove combining with unknown change. You may decide to pull together small groups of students to help them work through a particular problem that is difficult for them. See p. 122 and p. 134 for things to consider as you observe students at work on story problems.

Set C, problem 1: Tara had 5 crayons. Her father gave her some more. Now she has 9 crayons. How many crayons did her father give her?

Set C, problem 3: 10 people were on a bus. Some more people got on. Now there are 15 people on the bus. How many people got on the bus?

5 people Came on the bus beceaus 10 PeopLe was on the bus a hd 5 more =15

Total of 20

- What strategies do students use to construct their sums? Do they seem to work randomly, choosing a number to start with and then continuing to add on other numbers until they reach 20? ("There's a 9. This card is a 5. Let's see... 10, 11, 12, 13, 14. I need more. What if I also added this 3?") Do they keep track of how many more they need to make 20? ("I'm up to 12, so I need to find something that makes 8.") Do they look for specific combinations that make 20, or two combinations of 10?
- How do students combine numbers? Do they count from 1 each time? Do they count on from one of the numbers? Do they use knowledge of number combinations? Do they use number combinations they know to find new combinations?
- How do students determine that the game is over? Do they keep trying combinations of remaining cards to make 20? Do they reason about the cards that remain? ("We have three 2's, a 7, and an 8 left. You can't make 20, because if you put them all together you get 21, and you can't take away 1 to get 20.")

Activity

Sharing Story Problem Strategies

Halfway through Session 12, announce the end of Choice Time and gather students to share their strategies for solving the problems in Set C that everyone was to complete. As you have been doing throughout the investigation, gather several strategies for each problem and record them on the board or chart paper.

Sessions 10, 11, and 12 Follow-Up

Story Problems Students may continue their work on Story Problems, Sets A–C, and Set D (Challenges), as appropriate.

Math Games Students play one of the games introduced in this unit. They might teach someone at home to play Total of 20 (variation C on Student Sheet 7, Total of 10). Or, they might play a new or more challenging variation of another game they have taken home during this unit: Collect 25¢ Together, Dot Addition, Counters in a Cup, or On and Off.

Extension

Close to 20 Student pairs play this card game with a deck of Number Cards. Each player is dealt five cards. For each round, each player uses three cards to make a number as close as possible to 20. A player's score for the round is the difference between the number made and 20. If players make 20 exactly, their score is 0. At the end of each round, players draw three replacement cards from the deck. The game is over when there are no more cards. The winner is the player with the *lowest* score at the end of the game.

Example round:

Nadia is dealt these cards:

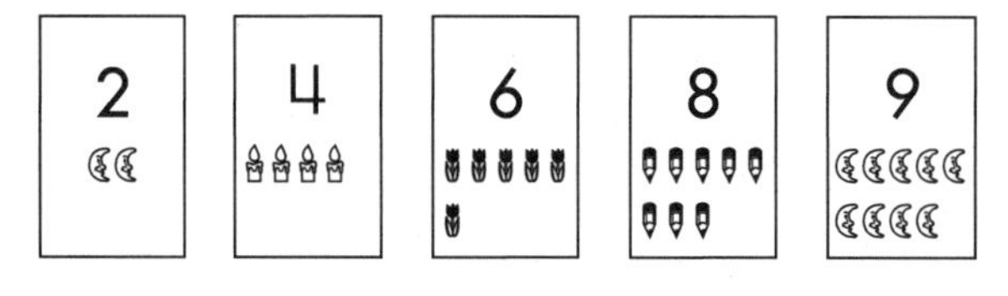

Claire is dealt these cards:

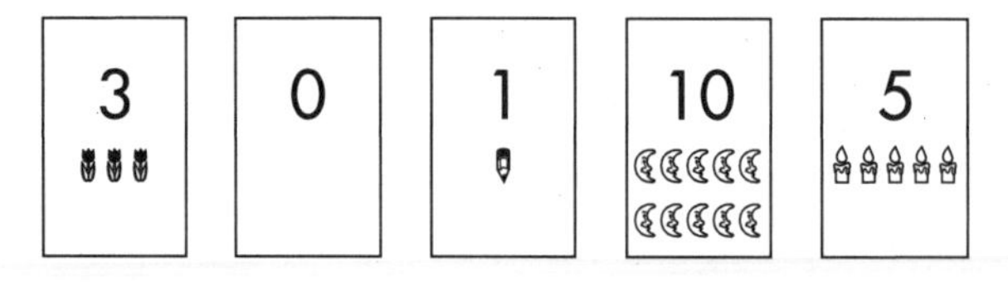

Nadia uses 8, 9, and 4 to make 21. Her score for the round is 1. Claire uses 10, 5, and 3 to make 18. Her score for the round is 2.

Session 13

Solving Story Problems

What Happens

As an assessment, students solve a variety of combining and separating story problems and record their solution strategies. Their work focuses on:

- solving combining and separating story problems
- recording strategies for solving combining and separating story problems, using pictures, numbers, words, and equations

Materials

- Story Problems, Set E, cut apart
- Unlined paper
- Paste or glue sticks
- Cubes or other counters (available)

Activity

Assessment

Solving Story Problems

Distribute Story Problems, Set E, one problem at a time. As before, students paste each problem on a sheet of paper, solve it, and record their solution strategies. Emphasize that you want students to find their own way to solve each problem and to record their strategies clearly so that you can see all their different methods.

The problem types in Set E are as follows:

- *Problem 1: Combining four amounts with unknown outcome.* The pencil jar has 8 red pencils, 3 blue pencils, 2 green pencils, and 7 yellow pencils. How many pencils are in the jar?
- *Problem 2: Separating with unknown outcome.* Ken had 18 pennies. He spent 7 of them. How many did he have left?
- *Problem 3: Combining with unknown change.* Tara drew 8 big stars. Then she drew some little stars. Now she has 14 stars on her paper. How many little stars did she draw?
- *Problem 4: Combining two amounts with unknown outcome (larger numbers than problem 1).* A frog hopped down 12 steps. Then it hopped down 12 more steps. How many steps did the frog hop down?
- *Problem 5: Separating with unknown outcome (larger numbers than problem 2).* Tara baked 30 muffins. Her friends ate 10 of them. How many were left?

Keep in mind that students who use numerical reasoning to solve problems with smaller numbers may rely on counting by 1's for problems with larger numbers. To see a good range of addition and subtraction strategies, ask all students to solve problems 1 and 2. They can then choose (or you might assign) one or more of the remaining problems as time permits. If many students had difficulty making sense of combining with unknown change, you may decide to leave out problem 3.

If the problems are too difficult for some students, adjust the numbers or substitute different problems. Have counters available for students to use.

Observing the Students

See the **Teacher Note,** Assessment: Solving Story Problems (p. 154), for more guidelines on assessing students' work.

- Are students comfortable with the combining and separating with unknown outcome problems? Do they know what they need to find in order to solve the problems? Can they keep the situation in mind as they solve them? Which students are comfortable with the combining with unknown change problem?
- What strategies do students rely on? Do they count out all the quantities in the problem? Do they start with one quantity and count on or back? Do they use number combinations they know to help solve the problem? Are students taking apart numbers in ways that help them solve problems more easily? Can they keep track of the parts of the numbers they create? Which students are beginning to use strategies involving tens and ones?
- Can students record their strategies in a way that makes sense, using some combination of pictures, words, numbers, and equations?

If there is time at the end of the session, call the class together to share a few strategies for one of the problems. Alternatively, students may continue their work on the Choice Time activities from Sessions 10–12.

Set E, problem 1: The pencil jar has 8 red pencils, 3 blue pencils, 2 green pencils, and 7 yellow pencils. How many pencils are in the jar?

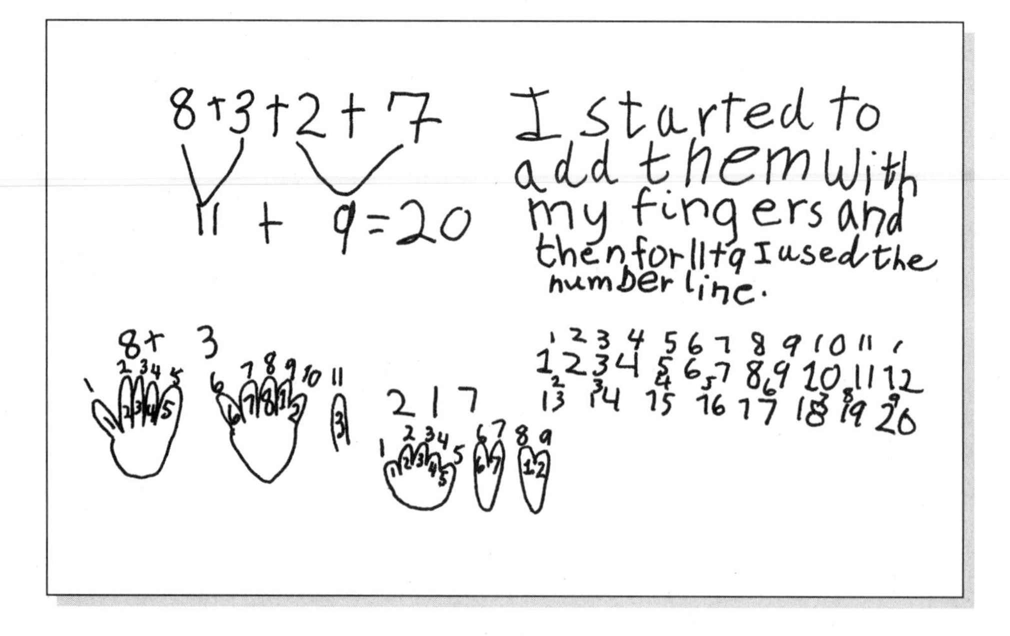

Set E, problem 2: Ken had 18 pennies. He spent 7 of them. How many did he have left?

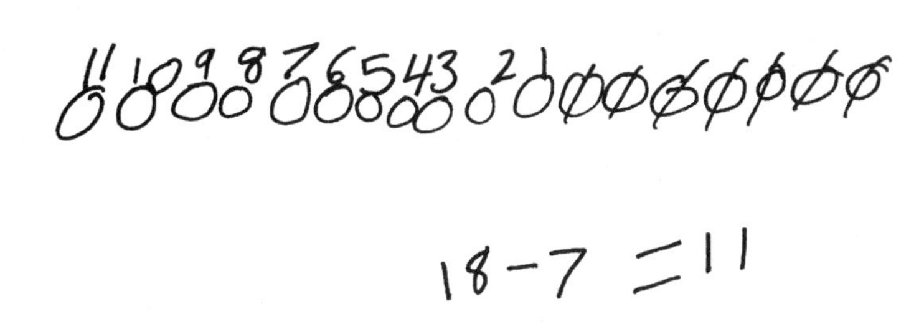

IKnow 8-7 so 18-7=11

1

Set E, problem 3: Tara drew 8 big stars. Then she drew some little stars. Now she has 14 stars on her paper. How many little stars did she draw?

I used my fangle
to ciet to 14 like Thes

1 2 3 4 5 6
8 9 10 11 12 13 14

8+6=14

Set E, problem 4: A frog hopped down 12 steps. Then it hopped down 12 more steps. How many steps did the frog hop down?

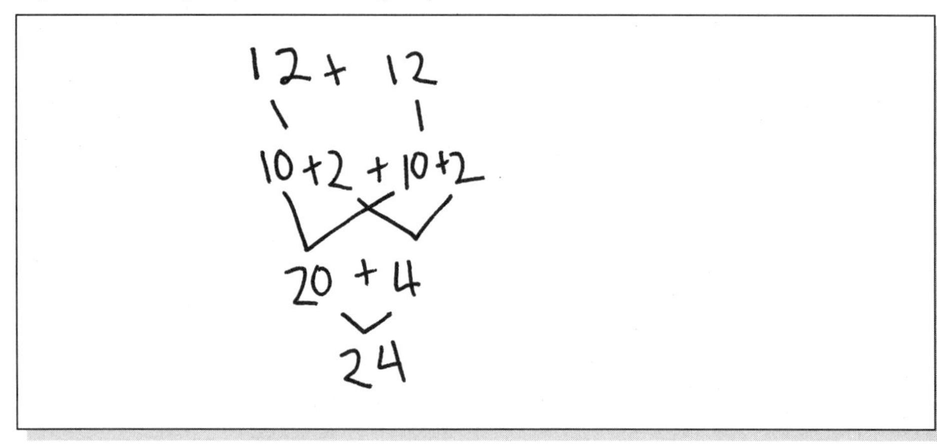

Set E, problem 5: Tara baked 30 muffins. Her friends ate 10 of them. How many were left?

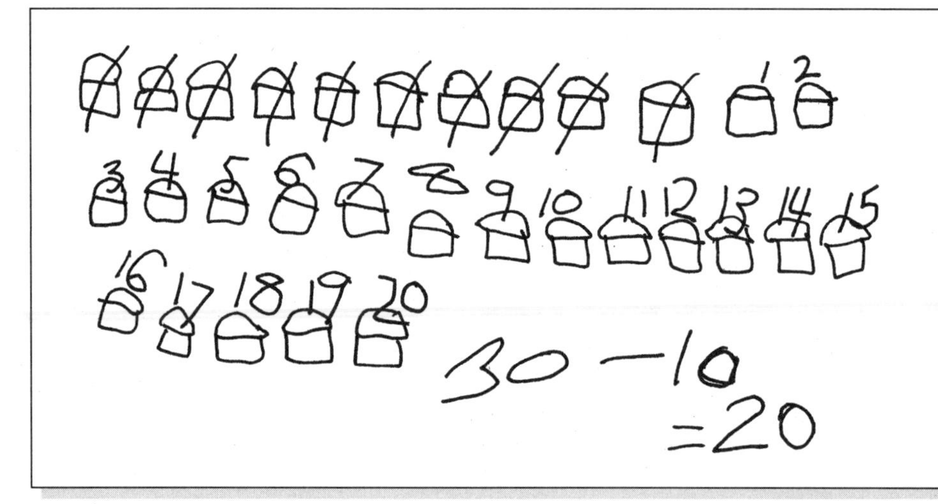

I cantde bak by 10's and I ende up with 20

Activity

Choosing Student Work to Save

As the unit ends, you may want to use one of the following options for creating a record of students' work on this unit.

- Students look back through their folders or notebooks and think about what they learned in this unit, what they remember most, what was hard or easy for them. Students might discuss this with a partner or share in the whole group.
- Depending on how you organize and collect student work, students might select some examples of their work to keep in a math portfolio. In addition you may choose some examples from each student's folder to include. Their work on Ten Crayons (p. 24), How Many Squares? (p. 91), and the Solving Story Problems activity (p. 149) can be useful pieces for assessing student growth over the school year. You may want to keep the original and make copies of these pieces for students to take home (or vice versa).
- Send a selection of work home for families to see. Some teachers include a short letter, summarizing the work in this unit. You could enlist the help of your students and together generate a letter that describes the mathematics they were involved in. This work should be returned to you if you are keeping a year-long portfolio of mathematics work for each student.

Teacher Note: Assessment: Solving Story Problems

As you observe students working on the final assessment activity, jot down notes about how each student is approaching the task, using the questions on pp. 150–152 as guidelines. Looking at these notes together with students' written work will give you a sense of students' strategies for solving combining and separating problems. You may find that students fall into three general groups, as follows.

In one group are students who are doing competent and appropriate work. These students can figure out what they need to do to solve combining and separating problems with unknown outcomes. They can select an appropriate strategy and record their work. They may use some strategies that involve number combinations, but rely primarily on strategies that involve counting by 1's. For example, for assessment problem 1 they may combine 8, 3, 2, and 7 by drawing a picture for each amount and counting how many pictures they have drawn. Or, they may know that 8 and 3 is 11, and then count up the rest, perhaps using their fingers to help them keep track.

For problem 2, they may take 7 from 18 by counting out 18 objects, removing 7, and counting the remaining set. Or, they may count down 7 from 18, perhaps using a list of numbers they have written or a number line to help them.

These students can solve problems with larger numbers (like problems 4 and 5), but will probably count out or draw all the quantities in the problem. Thus they may work slowly and may make an occasional counting error. They may need some help making sense of problems that involve combining with unknown change, or they may be able to solve these problems using strategies that involve counting by 1's.

In another group are students who, in addition to what the first group can do, often use strategies that involve number combinations or tens and ones. For problem 1, they might combine 8, 3, 2, and 7 by grouping the numbers in the problem to make two combinations of 10 (8 + 2 and 3 + 7), and then reason that 10 + 10 is 20. Or, they may explain that "8 and 3 is 11, and 2 and 7 is 9, then break 11 into 10 and 1, add the 1 and 9 to get 10, and the answer is 20 because 10 and 10 is 20."

For problem 2, they might take 7 from 18 by breaking up the numbers in the problem and recombining them in different ways: "18 is 8 and 10, and 8 take away 7 is 1. Then, 10 more is 11." Or, they might reason that since 7 from 8 is 1, 7 from 18 must be 10 more than 1, or 11.

These students may approach problems 4 and 5 with strategies that involve counting by 1's. Or, they might use strategies that involve number combinations or tens and ones. For example, in problem 4, they might combine 12 and 12 by breaking each 12 into 10 and 2, and explain that since the two 10's are 20 and the two 2's are 4, the total is 24. In problem 5, they might take 10 from 30 by explaining that it's 20 because when you count by 10's back from 30, you say 30, **20**, 10.

These students are probably beginning to make sense of problems that involve combining with unknown change; they may approach these problems with strategies that involve counting by 1's or numerical reasoning.

Continued on next page

Teacher Note

continued

In a third group are students about whom you have some concerns. They may have some difficulty understanding how to approach problems that involve separating, and they may have trouble explaining and recording their strategies. These students may not be comfortable working with numbers in the upper teens or 20's. They may have trouble keeping track of their counts in this range, especially when counting backwards, and they may not be able to count out 15 or more objects accurately. These students may have difficulty making sense of problems that involve combining with unknown change without some support from you.

These groupings can help you think about the class as a whole and about what individual students may need. Throughout the remainder of the year, students will benefit from opportunities to continue developing their addition and subtraction strategies. You might give students more work on particular kinds of story problems for homework, or as part of their work in subsequent Choice Times. You might also provide opportunities for students to share their strategies for solving some of these problems.

Teacher Note — About Choice Time

Choice Time is an opportunity for students to work on a variety of activities that focus on similar mathematical content. Choice Times are found in every unit of the grade 1 *Investigations* curriculum. These generally alternate with whole-class activities in which students work individually or in pairs on one or two problems. Each format offers different learning experiences; both are important for students.

In Choice Time the activities are not sequential; as students move among them, they continually revisit some of the important concepts and ideas they are learning. Many Choice Time activities are designed with the intent that students will work on them more than once. As they play a game a second or third time, or as they work to solve similar problems, students are able to refine their strategies, see a variety of approaches, and bring new knowledge to familiar experiences.

You may want to limit the number of students working on a particular Choice Time activity at any one time. In many cases, the quantity of materials available limits the number. Even if this is not the case, limiting the number is advisable because it gives students the opportunity to work in smaller groups. It also gives them a chance to do some choices more than one time. Often when a new choice is introduced, many students want to do it first. Assure them that, even with your limits, they will have the chance to try each choice.

Initially you may need to help students plan what they do. Rather than organizing them into groups and circulating the groups every 15 minutes, support students in making their own decisions. Making choices, planning their time, and taking responsibility for their own learning are important aspects of a student's school experience. If some students return to the same activity over and over again without trying other choices, suggest that they make a different first choice and then do the favorite activity as a second choice.

How to Set Up Choices

Some teachers prefer to have the choices set up at centers or stations around the room. At each center students will find the materials needed to complete the activity. Other teachers prefer to have materials stored in a central location, with students taking the materials to their own desks or tables. In either case, materials should be readily accessible, and students should be expected to take responsibility for cleaning up and returning materials to their appropriate storage locations. Giving a "5 minutes until cleanup" warning before the end of any session allows students to finish what they are working on and prepare for the upcoming transition.

Decide which arrangement to use in your classroom. You may need to experiment with a few different structures before finding the setup that works best for you and your students.

The Role of the Student

Establish clear guidelines when you introduce Choice Time. Discuss students' responsibilities:

- Try every choice at least once.
- Work with a partner or alone. (Some activities require that students work in pairs, while others can be done either alone or with a partner.)
- Keep track, on paper, of the choices you have worked on.
- Keep all your work in your math folder.
- Ask questions of other students when you don't understand or feel stuck. (Some teachers establish the rule, "Ask two other students before me," requiring students to check with two peers before coming to the teacher for help.)

For each Choice Time, list the activity choices on a chart, the board, or the overhead. Sketch a picture with each choice for students who may have difficulty reading the activity names. Some teachers laminate a piece of tagboard to create a Choices board that they can easily update as new choices are added from session to session and old choices are no longer offered.

First grade students can keep track of the choices they have completed in one of these ways:

- When they have completed an activity, students record its name or picture on a blank sheet of paper.
- Post a sheet of lined paper at each station, or a sheet for each choice at the front of the room. At the top of each sheet, put the name of one activity and the corresponding picture. When students have completed an activity, they print their name on the corresponding sheet. Keep these lists throughout an investigation, as the same choices may be offered several times.

Some teachers keep a date stamp at each Choice Time station or at the front of the room, making it easy for students to record the date as well.

In any classroom there will be a range of how much work students complete. Some choices include extensions and additional problems for students who have completed their required work. Encourage students to return to choices they have done before, do another problem or two from the choice, or play a game again. You may also want to make the choices available at other times during the day.

Whenever students do any work on paper during Choice Time, they put this in their math folders at the end of the session.

At the end of a Choice Time session, spend a few minutes discussing with students what went smoothly, what sorts of issues arose and how they were resolved, and what students enjoyed or found difficult about Choice Time. Having students share the work they have been doing often sparks interest in an activity. Some days, you might ask two or three volunteers to talk about their work. On other days, you might pose a question that someone asked you during Choice Time, so that other students might respond to it. Encourage students to be involved in the process of finding solutions to problems that come up in the classroom. In doing so, they take some responsibility for their own behavior and become involved with establishing classroom policies.

The Role of the Teacher

Choice Time provides you with the opportunity to observe and listen to students while they work. At times, you may want to meet with individual students, pairs, or small groups who need help. This gives you the chance to focus on students you haven't had a chance to observe before, or to do individual assessments. Recording your observations of students will help keep you aware of how they are interacting with materials and solving problems. The **Teacher Note,** Keeping Track of Students' Work (p. 158), offers some strategies for recording and using your observations.

During the initial weeks of Choice Time, much of your time will probably be spent in classroom management, circulating around the room, helping students get settled into activities, and monitoring the process of moving from one choice to another. Once routines are familiar and well established, students will become more independent and responsible for their work during Choice Time. This will allow you to spend more concentrated periods of time observing the class as a whole or working with individuals and small groups.

Teacher Note — Keeping Track of Students' Work

Throughout the *Investigations* curriculum, there are numerous opportunities to observe students as they work. Teacher observations are an important part of ongoing assessment. A single observation is like a snapshot of a student's experience with a particular activity, but when considered over time, a collection of these snapshots provides an informative and detailed picture of a student. Such observations can be useful in documenting and assessing student's growth, as well as in planning curriculum. They offer important sources of information when preparing for parent conferences or writing student reports.

The way you observe students will vary throughout the year. At times you may be interested in particular problem-solving strategies that students are developing. Other times, you might want to observe how students use or do not use materials for solving problems. You may want to focus on how students interact when working in pairs or groups. Or you may be interested in noting the strategy that a student uses when playing a game during Choice Time. Class discussions also provide many opportunities to take note of student ideas and thinking.

You will probably need some sort of system to record and keep track of your observations. While a few ideas and suggestions are offered here, it's important to find a record-keeping system that works for you. All too often, keeping observation notes on a class of 28–32 students can quickly become overwhelming and time-consuming.

A class list of names is one convenient way of jotting down your observations. Since the space is somewhat limited, it is not possible to write lengthy notes; however, over time, these short observations provide important information.

Another common approach is to keep a supply of adhesive address labels on clipboards around the room. After taking notes on individual students, you can peel off each label and stick it in the appropriate student's file.

Some teachers keep a loose-leaf notebook with a page for each student. When something about a student's thinking strikes them as important, they jot down brief notes and the date.

You may find that writing notes at the end of each week works well for you. Some teachers find this a useful way of reflecting on individual students, on the curriculum, and on the class as a whole. Planning for the next week's activities often grows out of these weekly reflections.

In addition to your own notes, you will have each student's folder of work for the unit. This documentation of their experiences can help you keep track of your students, assess their growth over time, and communicate this information to others. An activity at the end of each unit, Choosing Student Work to Save, suggests particular pieces of work you might keep in a portfolio of work for the year.

Strategies for Learning Addition Combinations

Teacher Note

In order to develop good computation strategies, students eventually need to become fluent with the *addition combinations* (number combinations with two addends, also called addition pairs) from 0 + 0 to 10 + 10. These combinations are part of the repertoire of number knowledge that contributes to the rich interconnections among numbers that we call *number sense.* A great deal of emphasis has been put on learning these combinations in elementary school. While we agree that knowing these combinations is important, we want to stress two important ideas:

- Students learn these combinations best by using strategies, not simply by rote memorization. Relying on memory alone is not sufficient, as many of us know from our own schooling. If you forget, as we all do at times, you are left with nothing. If, on the other hand, your learning is based on your understanding of numbers and their relationships, you have a way to rethink and restructure your knowledge when you don't remember something you thought you "knew."
- Knowing the combinations should be judged by fluency in use, not necessarily by instantaneous recall. Through repeated use and familiarity, students will eventually come to know most of the addition combinations immediately, and a few by using some quick and comfortable numerical reasoning strategy. For example, when one of the *Investigations* authors thinks of 8 + 5, she doesn't automatically see the total as 13; rather, she sees the 5 broken apart into a 2 and a 3, the 2 combined with the 8 to make 10, then the 10 and the 3 combined to total 13. While this strategy takes quite a while to write down or to read, she "sees" this relationship almost instantaneously. As far as she is concerned, she "knows" this combination.

We avoid calling these addition combinations the addition "facts," because we think the term "facts" tends to elevate knowledge of these combinations above other mathematical knowledge, as if knowing these is the most important thing in mathematics. Developing fluency in using these combinations is important, but many other ideas are just as critical for building number sense.

Using Strategies to Learn Addition Combinations

At grade 1, many students are still primarily figuring out number combinations by counting by 1's. They are not yet thinking about larger units, so they are not yet ready to use strategies to learn number combinations. However, if you see some students begin to develop strategies based on number relationships, you can encourage them to articulate and build on this thinking. Students at this age begin to use several important strategies:

- **Adding 1's and 2's** Adding 1 or 2 easily is a beginning step in learning addition combinations. Students begin to "just know" that any time you add 1 to a number, you end up with the next number in the counting sequence. After students have had many experiences adding on one object to a group of objects, they are able to coordinate the idea of adding on 1 with their knowledge of the counting numbers. Building on this idea, they begin to add on 2 by quickly counting on two more numbers.
- **The Doubles** Students learn many of the doubles (3 + 3, 4 + 4, 5 + 5, and so forth) quite early. Some students can begin to use the doubles they know to help them figure out other combinations: "I know that 5 + 5 is 10, so 6 + 5 is one more, that's 11."

Continued on next page

Teacher Note

continued

- **Sums That Make 10** Some students may begin to recognize some of the combinations that make 10 and can eventually build on these to find others. For example, knowing that 6 + 4 is 10, students may be able to imagine taking 1 from the 4 and adding it to the 6, giving them 7 + 3.

Encourage your students to share their strategies for learning number combinations and to explain how they figured out new or difficult combinations. The **Teacher Note**, Building on Number Combinations You Know (p. 22), discusses strategies you might see your students using in this investigation. Keep in mind, though, that first grade students are just becoming familiar with number combinations. By the end of first grade, most will be comfortable with combinations that involve +1 and +2, and many will also know quite a few other combinations, including several doubles and combinations of 10. But many students will not yet be able to use the combinations they know to find other combinations, and most will not yet recognize that, for example, adding 2 + 8 is the same as adding 8 + 2. Your students will have many opportunities to continue work on number combinations in second grade and the early part of third grade.

Writing and Recording

Just as students should be engaged in frequent mathematical conversation, so too should they be encouraged to explain their problem-solving strategies in writing and with pictures and diagrams. Writing about how they solved a problem is a challenging task, but one that is worth the investment of time. As with any writing assignment, many students will need support and encouragement as they begin to find ways of communicating their ideas and thinking on paper, but even the very youngest students can be encouraged to represent their problem-solving strategies.

The range of students' abilities to write and record will vary greatly in any first grade classroom. First graders are just becoming familiar with the areas of reading and writing, and reflecting on one's own thinking is a challenging task in any case. Initially, some students will record just a few words, pictures, or numbers that describe their strategy. Encouraging students to draw pictures as part of their explanation is often a way into the task for many students. Explain to students that mathematicians often write and draw about their ideas as a way of explaining to others how they are thinking about a problem. The more often students are encouraged (and expected) to write and record their ideas, the more comfortable and fluent they become.

Students benefit tremendously from discussing their ideas before writing about them. Sometimes this might happen in pairs, other times in whole-group discussions. Questions and prompts such as "How did you solve the problem?" or "Can you tell me what you did after you put the cubes into groups of five?" may help extend students' thinking. During whole-class discussion, you will model writing and recording strategies for students so they can see ways to record their mathematical strategies using words, numbers, and pictures. For example, when one boy reported to the class how he solved a problem about combining 6 and 5, he said: "I counted out 5 cubes. I counted out 6 more. Then I counted the cubes, and it stopped at 11."

The teacher recorded his strategy on the board as follows, reiterating his strategy in words as she recorded:

I counted out 5 cubes. I counted out 6 cubes.

■ ■ ■ ■ ■ □ □ □ □ □ □

1 2 3 4 5 6 7 8 9 10 11

I counted all the cubes.

As with any type of writing, providing feedback to students is an important part of the process. As the audience for your students' work, you can point out those ideas that clearly convey their thinking and those that need more detail. If students read their work aloud to you or to a classmate, quite often they can identify by themselves ideas that are unclear and parts that are incomplete. See the **Dialogue Box,** Helping Students Record Their Strategies (p. 57), for one teacher's interactions with students.

Teacher Note — *Introducing Notation*

While first graders should become familiar with standard equation notation, it is not essential that they use it themselves at this level. The equation format may seem very straightforward to us as adults, but it actually assumes some complex ideas about number relationships. Second graders are generally more ready to use equation notation to record their work than are first graders. In first grade, many students are grappling with the problem situations and deciding what actions are required to solve them. They are moving from counting by 1's in all situations to sometimes thinking of numbers in larger chunks. They are just becoming familiar with some number combinations. The emphasis for students must remain firmly on making sense of the problem situation and finding a way to solve it. This emphasis includes recording solutions in a way that comes from the student, rather than from the teacher.

As you use equations correctly, students can see how they are used, and some will incorporate them in their own recording. Introduce equations naturally as a way to record student strategies when these strategies are numerical ones. For example, suppose one of your students describes this approach to combining 4 and 8: "I knew that 2 and 2 is 4, so I said 8 and 2 is 10, and then I added the 2 more, and 10 and 2 is 12." This strategy can be nicely recorded with equations:

$2 + 2 = 4$

$8 + 2 = 10$

$10 + 2 = 12$

However, if a student says, "I counted up from 8, so that was 9, 10, 11, 12," this strategy does not suggest equation format. It would be better recorded something like this:

8 ● ● ● ●
9 10 11 12

Try to match your way of recording as best you can to what the students say, so that a variety of ways of thinking about and recording the problem are valued and shared. At grade 1, it is more important for students to develop clear ways of thinking about the problem for themselves than to use standard notation.

When you do use equations to record, be sure you use them correctly so that students will have a good model. In particular, use *separate* equations to model consecutive steps in a student's thinking:

$8 + 2 = 10$

$10 + 2 = 12$

It is incorrect to write $8 + 2 = 10 + 2 = 12$, which means that the sum of $8 + 2$ is equivalent to the sum of $10 + 2$. Of course, what a first grader means by this is clear: that she added 8 and 2 first, then added 2 to their sum to get 12. The first grader who writes this way simply understands the notation as a sequence of events, rather than as an equation. Likewise, some first graders will not recognize that "order matters" when using subtraction notation. They might record $2 - 9$ to show that they took 2 away from 9. Encourage students to think about whether $2 - 9$ means the same thing as $9 - 2$.

By the end of second grade, students can be guided to use equation notation correctly, but first grade is too early to do that for most students.

Using the Calculator in First Grade

Teacher Note

Increasingly sophisticated calculators are used everywhere, from homes to high school classrooms to the workplace. If students are forbidden to use calculators in school while they see adults using them outside of school, they learn that "school" mathematics is nothing like mathematics in the real world. In the world around them, using a calculator is part of real life. We believe that students at all levels need to learn how to use calculators effectively and appropriately as a tool, just as throughout the elementary grades they learn to interpret maps, measure with rulers, and use coins.

Calculators enable students at all levels to apply their reasoning and problem-solving skills to a wider variety of problems. Students can combine the use of calculators with mental calculation or work with manipulatives as they solve problems that use large numbers or require many calculations.

For example, in the middle of the year in one first grade class, students were asked to determine how many crackers the teacher had brought in for snack: there were eight bags of crackers, and each contained six crackers. Of the many methods students came up with, several involved combining the use of the calculator with other mathematical tools.

- One student recorded 6 + 6 + 6 + 6 + 6 + 6 + 6 + 6, explaining that each 6 represents the number of crackers in one bag, and then used a calculator to find the sum.
- Another student also began by recording 6 + 6 + 6 + 6 + 6 + 6 + 6 + 6, but was able to solve more of the problem in her head. She knew that 6 + 6 is 12, and recorded 12 under each pair of sixes:

  6 + 6 + 6 + 6 + 6 + 6 + 6 + 6

  \/ \/ \/ \/

  12 + 12 + 12 + 12

  She then used the calculator to add the four twelves.
- Another student used the calculator to help her solve the problem by successive "doubling." She explained that since there are six in one bag, there are 12 in two bags. She knew that doubling 12 would give her the number in four bags, and doubling the number in four bags would give her the total in all eight. She then used the calculator to add 12 and 12 to get 24, and then to add 24 and 24 to get 48.
- Yet another student recognized the problem as a multiplication situation. He knew that the problem could be represented by 6 × 8, but did not know how much that was. He used the calculator to find 6 × 8. Then, to check, he built a representation with eight towers of six interlocking cubes, and counted all the cubes.

When Should Students Use Calculators? You will probably find many situations in which using calculators can facilitate students' work with large numbers, as it did for the bags of crackers problem. Some of these situations may arise during math time, others outside of math (for example, finding the total number of cans collected for a recycling project, the amount of money collected at a bake sale, the number of cookies needed for a class party).

You will find also many situations in which you do *not* want students to use calculators. When students are exploring number combinations with the How Many of Each? problems, and when they are developing strategies for solving story problems, we suggest that students focus on strategies built on what they know about counting and number relationships.

As students progress through the elementary grades, they will begin making their own choices about when it is appropriate to combine calculators with their own reasoning and when it is appropriate to use other tools, such as mental calculation and estimation, paper and pencil, and manipulatives. Students should have opportunities throughout grade 1 to freely explore and experiment with all these tools.

Teacher Note — Creating Your Own Story Problems

The story problems provided in this unit are combining and separating situations. Many teachers like to create their own problems; this enables them to use story contexts that reflect the interests, knowledge, and environment of their own students, as well as to adjust the numbers appropriately. We expect and encourage this; however, whenever you change the contexts or numbers in a problem from this investigation, avoid altering its underlying structure.

Creating Interesting Contexts Use contexts that are interesting to students without being distracting. Teachers find that simple situations, familiar to all their students, are the most satisfying. One source of good situations is experiences that you know all your students have had. For example, one urban class walks to a nearby park every day for recess; a problem based on that experience might be:

> Today at the park, I counted 6 squirrels on the ground and 2 more in a tree. How many squirrels did I see in the park?

Another class may be having bake sales at lunchtime to raise money for a class trip:

> On Friday we sold 6 chocolate cupcakes and 2 vanilla cupcakes. How many cupcakes did we sell?

In one classroom the teacher made up two characters like her students, Ted and Sophia, and built problem situations around them:

> Ted and Sophia got a bag of peanuts. As they walked home, Ted ate 6 peanuts and Sophia ate 2 peanuts. How many peanuts did they eat?

> Sophia and Ted went to the post office. Ted bought 2 stamps and Sophia bought 6 stamps. How many stamps did they buy?

Many teachers also take advantage of special events, classroom happenings, seasons, or holidays for problem contexts:

> Kira made 6 snowballs and Jake made 2. How many snowballs did they have?

Related Problems Simple, familiar situations often suggest other problems that easily follow from the initial one. Such follow-up questions might be provided to everyone as options, or only to students who have finished the first problem. Additional problems related to the previous examples might include these:

> Yesterday I counted 7 squirrels on the ground and 3 in the trees. Did I see more squirrels yesterday or today?

> If we charged 2¢ for each cupcake, how much did we get for all the cupcakes?

> If we sold 10 more cupcakes, how many cupcakes would we sell in all?

> If Kira and Jake want to have 11 snowballs, how many more do they need to make?

When creating follow-up problems, consider whether to keep the level of challenge the same or to make the problem easier or more difficult.

Adjusting Numbers in the Problem You will probably have students who are comfortable working with numbers in the upper teens and higher, and other students who need to work with smaller numbers. In some classrooms, teachers choose two sets of numbers for each problem. For example:

> Sophia has 5 pennies. She earned 6 cents. How many pennies does she have now?

> Sophia has 12 pennies. She earned 7 cents. How many pennies does she have now?

Understanding addition and subtraction at this level involves making sense of story problems, having strategies for solving them, and being able to communicate strategies orally and in writing. While you may sometimes want to challenge students by giving them larger numbers, you can also challenge them to find answers in more than one way and to find clearer ways to explain their work.

"Key Words": A Misleading Strategy

Teacher Note

Some mathematics materials have advocated a "key words" technique to help students solve story problems. Students are taught to recognize words in a problem that provide clues about how to choose which operation to use to solve it. For example, *altogether* or *more* are said to signal addition, whereas *left* or *fewer* signal subtraction.

> I had 5 marbles. Lee gave me 6 more. How many do I have *altogether?*
>
> I had 16 marbles. I gave away 8. How many do I have *left?*
>
> I had 16 marbles. Lee has 7 *fewer* than I do. How many marbles does Lee have?

There are two flaws in the key word approach. First, these words may be used in many ways. They might be part of a problem that requires a different operation from the expected one:

> There are 28 students in our class altogether. There are 13 boys. How many are girls?

If we trust in key words, then *altogether* in this problem should signal addition of the numbers in the problem, 28 + 13. In fact, the problem calls for finding the *difference* between 28 and 13.

The second reason for avoiding reliance on key words is that students should think through the entire structure of the problem. They need to read the problem and understand the situation so that they can construct a model of the problem for themselves. Here's another example:

> I am making cookies for my party. There will be 6 people at my party, including me. I want each person to have 4 cookies. How many cookies should I make altogether?

If students are encouraged to use key words, they are likely to simply pull numbers out of the problem and carry out some operation (in this case, perhaps, 6 + 4) without developing a model of the whole problem (in this case, 6 equal groups of 4).

Counting

Counting is an important focus in the grade 1 *Investigations* curriculum, as it provides the basis for much of mathematical understanding. As students count, they are learning how our number system is constructed, and they are building the knowledge they need to begin to solve numerical problems. They are also developing critical understandings about how numbers are related to each other and how the counting sequence is related to the quantities they are counting.

Counting routines can be used to support and extend the counting work that students do in the *Investigations* curriculum. As students work with counting routines, they gain regular practice with counting in familiar classroom contexts, as they use counting to describe the quantities in their environment and to solve problems based on situations that arise throughout the school day.

How Many Are Here Today?

Since you must take attendance every day, this is a good time to look at the number of students in the classroom in a variety of ways.

Ask students to look around and make an estimate of how many are here today. Then ask them to count.

At the beginning of the year, students will probably find the number at school today by counting each student present. To help them think about ways to count accurately, you can ask questions like these:

How do we know we counted accurately? What are different methods we could use to keep track and make sure we have an exact count? (For example, you could count around a circle of seated students, with each student in turn saying the next number. Or, all students could start by standing up, then sit down in turn as each says the next number.)

Is there another way we could count to double-check? (For example, if you counted around the circle one way, you could count around the circle the other way. If you are using the standing up/sitting down method, you could recount in a different order.)

You might want to count at other times of the day, too, especially when several students are out of the room. For example, suppose groups of students are called to the nurse's office for hearing examinations. Each time a new group of students leaves, you might ask the class to look around and think about how many students are in the room now:

So, this time Diego's table and Mia's table both went to the nurse. Usually we have 28 students here. Look around. What do you think? Don't count. Just tell me about how many students might be here now. Do you think there are more than 5? more than 10? more than 20?

Later in the year, some students may be able to use some of the information they know about the total number of students in the class and how many students are absent to reason about the number present. For example, suppose 26 students are in class on Monday, with 2 students absent. On Tuesday, one of those students comes back to school. How many students are in class today? Some students may still not be sure without counting from one, but other students may be able to reason by counting on or counting back, comparing yesterday and today. For example, a student might solve the problem in this way:

> "Yesterday we had 26 students, and Michelle and Chris were both absent. Today, Chris came back, so we have one more person, so there must be 27 today."

Another might solve it this way:

> "Well we have 28 students in our class when everyone's here. Now only Michelle is absent, so it's one less. So it's 27."

From time to time, you might keep a chart of attendance over a week or so, as shown below. This helps students become familiar with different combinations of numbers that make the same total. If you have been doing any graphing, you might want to present the information in graph form.

Day	Date	Present	Absent	Total
Monday	March 2	26	2	28
Tuesday	March 3	27	1	28
Wednesday	March 4	27	1	28
Thursday	March 5	27	1	28
Friday	March 6	28	0	28
Monday	March 9	28	0	28
Tuesday	March 10	26	2	28
Wednesday	March 11	25	3	28

After a week or two, look back over the data you have collected. Ask questions about how things have changed over time.

In two weeks of attendance data, what changes? What stays the same?

On which day were the most students here? How can you tell? Which day shows the least students here? What part of the [chart] gives you that information?

Another idea (for work with smaller numbers) is to keep track of the number of girls and boys present and absent each day. Again, many students will count by 1's. Later in the year, some will also reason about these numbers:

> "There are two people absent today and they're both girls. We usually have 14 girls, and Kaneisha's sick, that's 13, and Claire's sick, that's 12."

Can Everyone Have a Partner?

Attendance can be an occasion for students to think about making groups of two:

We have 26 students here today. Do you think that everyone can have a partner if we have 26 students?

Students can come up with different strategies for solving this problem. They might draw 26 stick figures, then circle them in 2's. They could count out 26 cubes, then put them together in pairs. They might arrange themselves in 2's, or count by 2's.

At the beginning of the year, many of your students will need to count by 1's from the beginning each time you add two more students, but gradually some will begin to notice which numbers can be broken up into pairs:

> "I know 13 doesn't work, because you can do it with 12, and 13's one more, so you can't do it."

Some students will begin to count by 2's, at least for the beginning of the counting sequence. Then, as the numbers get higher, they may still be able to keep track of the 2's, but need to count by 1's:

> "So, that's 2, 4, 6, 8, 10, 12, um, 13, 14 . . . 15, 16."

As you explore 2's with your students, keep in mind that many of them will need to return to 1's as a way to be sure. Even though some students learn the counting sequence 2, 4, 6, 8, 10, 12 . . . by rote, they may not connect this counting sequence to the quantities it represents at each step.

One teacher found a way to help students develop meaning for counting by 2's. She took photographs of each student, backed them with cardboard, then used them during the morning meeting as a model for making pairs. She laid out the photos in two columns, and asked about the new total after the addition of each pair:

We have 10 photos out so far. The next two photos are for William and Yanni. When we put those two photos down, how many photos will we have?

Lining up is another time to explore making pairs. Before lining up, count how many students are in class (especially if it's different from when you took attendance). Ask students whether they think the class will be able to line up in even pairs. For many first grade students, the whole class is too many people to think about. You can ask about smaller groups:

What if Kristi Ann's table lines up first? Do you think we could make even partners with the people at that table?

What about Shavonne's table? ... Do you think Shavonne's table will have an extra? How do you know?

Is there another table that would have an extra that we could match up with the extra person from Shavonne's table?

Once students are lined up in pairs, they can count off by 2's. Because most first graders will need to hear all the numbers to keep track of how the counting matches the number of people, ask them to say the first number in the pair softly and the second one loudly. Thus the first pair in line can say, "1, **2**," the second pair can say, "3, **4**," and so forth.

Counting to Solve Problems

Be alert to classroom activities that lend themselves to a regular focus on solving problems through counting. Use these situations as contexts for counting and keeping track, estimating small quantities, breaking quantities into parts, and solving problems by counting up or back. For example, take a daily milk count:

Everyone who is buying milk today stand up. Without counting yet, who has an idea how many students might be standing up? Is it more than 5? more than 10? more than 50? ... Now, let's count. How could we keep track today so that we get an accurate count?

You can make a problem out of lunch count:

We found out that 23 students are buying school lunch today. We have 27 students here. So how many students brought their own lunch from home today?

Watch for the occasional sharing situation:

Claire brought in some cookies she made to share for snack. She brought 36 cookies. Is that enough for everyone to have one cookie, including me and our student teacher? Oh, and Claire wants to invite her little brother to snack. Do we have enough for him, too? Will there be any cookies left over?

The sharing of curriculum materials can also be the basis of a problem:

Each pair of students needs a deck of number cards to share. While I'm getting things together, work on this problem with your partner. We said this morning that we have 26 students here. If I need one deck for each pair, how many decks do I need?

Exploring Data

Through data routines at grade 1, students gain experience working with categorical data—information that falls into categories based on a common feature (for example, a color, a shape, or a shared function). The data routines specifically extend work students do in the *Investigations* curriculum. The Guess My Rule game and its many variations (introduced in the unit *Survey Questions and Secret Rules*) can be used throughout the year for practice with organizing sets into categories and finding ways to describe those categories—a fundamental part of analyzing data. Students can also practice collecting and organizing categorical data with quick class surveys that focus on their everyday experiences; this practice supports the survey-taking they do in the curriculum.

Guess My Rule

Guess My Rule is a classification game in which players try to figure out the common characteristic, or attribute, of a set of objects. To play the game, the rule maker (who may be the teacher, a student, or a small group) decides on a secret rule for classifying a particular group of things. For example, a rule for classifying people might be WEARING STRIPES.

The rule maker (always the teacher when the game is first being introduced) starts the game by giving some examples of objects or people who fit the rule. The guessers then try to find other items that fit the same rule. Each item (or person) guessed is added to one of two groups—either *does fit* or *does not fit* the rule. Both groups must remain clearly visible to the guessers so they can make use of all the evidence as they try to figure out the rule.

Emphasize to the players that "wrong" guesses are as important as "right" guesses because they provide useful clues for finding the rule. When you think most students know the rule, ask for volunteers to share their ideas with the class.

Once your class is comfortable with the activity, students can choose the rules. Initially, you may need to help students choose appropriate rules.

Guess my Rule with People When sorting people according to a secret rule, always base the rule on just one feature that is clearly visible, such as WEARING A SHIRT WITH BUTTONS, or WEARING BLUE. When students are choosing the rule, they may choose rules that are too obvious (such as BOY/GIRL), so vague as to apply to nearly everyone (WEARING DIFFERENT COLORS), or too obscure (HAS AN UNTIED SHOELACE). Guide and support students in choosing rules that work.

Guess My Rule with Objects Class sets of attribute blocks (blocks with particular variations in size, shape, color, and thickness) are a natural choice for Guess My Rule. You can also use collections of objects, such as sets of keys, household container lids, or buttons. One student sorts four to eight objects according to a secret rule. Others take turns choosing an object from the collection that they think fits the rule and placing it in the appropriate group. If the object does not fit, the rule maker moves it to the NOT group. After several objects have been correctly placed, students can begin guessing the rule.

Guess My Object Once students are familiar with Guess My Rule, they can use the categories they have been identifying to play another guessing game that also involves thinking about attributes. In this routine, students guess, by the process of elimination, which particular one of a set of objects has been secretly chosen. This works well with attribute blocks or object collections.

To start, place about 20 objects where everyone can see them. The chooser secretly selects one of the objects on display, but does not tell which one (you may want the chooser to tell you, privately). Other students ask yes-or-no questions, based on attributes, to get clues to help them identify the chosen object. After each answer, students move to one side the objects that have been eliminated. That is, if someone asks "Is it round?" and the answer is yes, all objects that are *not* round are moved aside.

Pause periodically to discuss which questions help eliminate the most objects. For example, "Is it this one?" eliminates only one object, whereas "Is it red?" may eliminate several objects. For more challenge, students can play with the goal of identifying the secret object with the fewest questions.

Quick Surveys

Class surveys can be particularly engaging when they connect to activities that arise as a regular part of the school day, and they can be used to help with class decisions. As students take surveys and analyze the results, they get good practice with collecting, representing, and interpreting categorical data.

Early in first grade, to keep the surveys quick and the routine short, use questions that have exactly two possible responses. For example:

Would you rather go outside or stay inside for recess today?

Will you drink milk with your lunch today?

Do you need left-handed or right-handed scissors?

As the school year progresses, you might include some survey questions that are likely to have more than two responses:

Which of these three books do you want me to read for story time?

Who was your teacher last year?

Which is your favorite vegetable growing in our class garden?

How old are you?

In which season were you born?

Try to choose questions with a predictable list of just a few responses. A question like "What is your favorite ice cream flavor?" may bring up such a wide range of responses that the resulting data is hard to organize and analyze.

As students become more familiar with classroom surveys, invite the class to brainstorm questions with you. You may decide to avoid survey questions about sensitive issues such as families, the body, or abilities, or you might decide to use surveys as a way of carefully raising some of these issues. In either case, it is best to avoid questions about material possessions ("Does your family have a car?").

Once the question is chosen, decide how to collect and represent data. Be sure to vary the approach. One time, you might collect data by recording students' responses on a class list. Another time, you might take a red interlocking cube for each student who makes one response, a blue cube for each student who makes the other response. Another time, you might draw pictures. If you have prepared Kid Pins and survey boards for use in *Mathematical Thinking at Grade 1,* these can be used for collecting the data from quick surveys all year.

Initially, you may need to help students organize the collected data, perhaps by stacking cubes into "bars" for a "graph," or by making a tally. Over time, students can take on more responsibility for collecting and organizing the data.

Always spend a little time asking students to describe, compare, and interpret the data.

What do you notice about these data?

Which group has the most? the least? How many more students want [recess indoors today]?

Why do you suppose more would rather [stay inside]? Do you think we'd get similar data if we collected on a different day? What if we did the same survey in another class?

Understanding Time and Changes

These routines help students develop an understanding of time-related ideas such as sequencing of events, understanding relationships among time periods, and identifying important times in their day.

Young students' understanding of time is often limited to their own direct experiences with how important events in time are related to each other. For example, explaining that an event will occur *after* a child's birthday or *before* an important holiday will help place that event in time for a child. Similarly, on a daily basis, it helps to relate an event to a benchmark time, such as *before* or *after* lunch. Both calendars and daily schedules are useful tools in sequencing events over time and preparing students for upcoming events. These routines help young students gain a sense of basic units of time and the passage of time.

Calendar

The calendar is a real-world tool that people use to keep track of time. As students work with the calendar, they become more familiar with the sequence of days, weeks, and months, and the relationships among these periods of time. Calendar activities can also help students become more familiar with relationships among the numbers 1–31.

Exploring the Monthly Calendar At the start of each month, post the monthly calendar and ask students what they notice about it. Some students might focus on arrangement of numbers or total number of days, while others might note special events marked on the calendar, or pictures or designs on the calendar. All these kinds of observations help students become familiar with time and ways that we keep track of time. You might record students' observations and post them near the calendar.

As the year progresses, encourage students to make comparisons between the months. Post the calendar for the new month next to the calendar for the month just ending and ask students to share their ideas about how the two calendars are similar and different.

Months and Years To help students see that months are part of a larger whole, display the entire calendar year on a large sheet of paper. Cut a small calendar into individual monthly pages and post the sequence of months on the wall. You might decide to post the months according to the school year, September through August, or the calendar year January through December. At the start of each month, ask students to find the position of the new month on the larger display. From time to time, you might also use this display to point out dates and distances between them as you discuss future events or as you discuss time periods that span a month or more. (Last week was February vacation. How many weeks until the next vacation?)

How Many More Days? Ask students to figure out how long until special events, such as birthdays, vacations, class trips, holidays, or future dates later in the month. For example:

Today is October 5. How many more days until October 15?

How many more days until [Nathan's] birthday?

How many more days until the end of the month?

Ask students to share their strategies for finding the number of days. Initially, many students will count each subsequent day. Later, some students may begin to find their answers by using their growing knowledge of calendar structure and number relationships:

> "I knew there were three more days in this row and I added them to the three days in the next row. That's 6 more days."

Others may begin using familiar numbers such as 5 or 10 in their counting:

> "Today is the 5th. Five more days is 10, and five more is 15. That's 10 more days until October 15."

For more challenge, ask for predictions that span two calendar months. For example, you might post the calendar for next month along side of the calendar of this month and ask a question like this:

It's April 29 today. How many more days until our class trip on May 6?

Note that we can refer to a date either as October 15 or as the 15th day of October. Vary the way you refer to dates so that students become comfortable with both forms. Saying "the 15th day of October" reinforces the idea that the calendar is a way to keep track of days in a month.

How Many Days Have Passed? Ask questions that focus on events that have already occurred:

How many days have passed since [a special event]? since the weekend? since vacation?

Mixed-Up Dates If your monthly display calendar has date cards that can be removed or rearranged, choose two or three dates and change their position on the calendar so that the numbers are out of order. Ask students to fix the calendar by pointing out which dates are out of order.

Groups of two or three can play this game with each other during free time. Students can also remove all the date cards, mix them up, and reassemble the calendar in the correct order. You might mark the space for the first day of the month so that students know where to begin.

Daily Schedule

The daily schedule narrows the focus of time to hours and shows students the order of familiar events over time. Working with schedules can be challenging for many first graders, but regular opportunities to think and talk about the idea will help them begin predicting what comes next in the schedule. They will also start to see relationships between particular events in the schedule and the day as a whole.

The School Day Post a schedule for each school day. Identify important events (start of school, math, music, recess, reading, lunch) using pictures or symbols and times. Include both analog (clock face) and digital (10:15) representations. Discuss the daily schedule each day with students using words such as *before* math, *after* recess, *during* the morning, *at the end of* the school day. Later in the school year you can begin to identify the times that events occur as a way of bridging the general idea of sequential events and the actual time of day.

The Weekend Day Students can create a daily schedule, similar to the class schedule, for their weekend days. Initially they might make a "timeline" of their day, putting events in sequential order. Later in the year they might make another schedule where they indicate the approximate time of day that events occur.

Weather

Keeping track of the weather engages young students in a real-life data collection experience in which the data they collect changes over time. By displaying this ongoing collection of data in one growing representation, students can compare changes in weather across days, weeks, and months, and observe trends in weather patterns, many of which correspond to the seasons of the year.

Monthly Weather Data With the students, choose a number of weather categories (which will depend on your climate); they might include sunny, cloudy, partly cloudy, rainy, windy, and snowy.

If you vary the type of representation you use to collect monthly data, students get a chance to see how similar information can be communicated in different ways. On the following page you'll see some ways of representing data that first grade teachers have used.

At the end of each month (and periodically throughout the month), ask questions to help students analyze the data they are collecting.

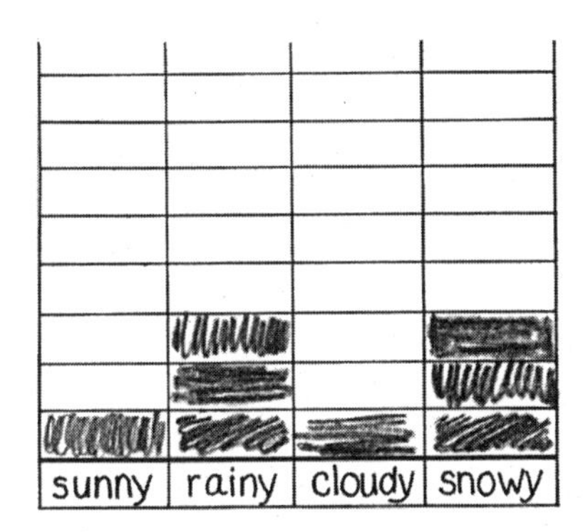

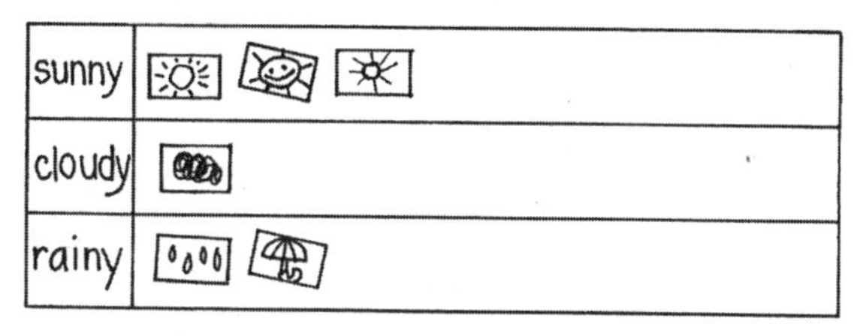

sun	clouds
rain	snow

Weather data can be collected on displays like these. In the second example, a student draws each day's weather on an index card to add to the graph. The third example uses stick-on dots.

What is this graph about?

What does this graph tell us about the weather this month (so far)?

What type of weather did we have for the most days? What type of weather did we hardly ever have?

How is the weather this month different from the weather last month? What are you looking at on the graph to help you figure that out?

How do you think the weather graph for next month will look?

Yearly Weather Data If you collect and analyze weather data for some period of time, consider extending this over the entire school year. Save your monthly weather graphs, and periodically look back to see and discuss the changes over longer periods of time.

Another approach over the entire year is to prepare 10-by-10 grids from 1-inch graph paper, making one grid for each weather category your class has chosen. Post the grids, labeled with the identifying weather word. Each day, a student records the weather by marking off one square on one or more grids; that is, on a sunny day, the student marks a square on the "sunny" grid, and if it's also windy, he or she marks the "windy" grid, too.

From time to time, students can calculate the total number of days in a certain category by counting the squares. Because these are arranged in a 10-by-10 grid, some students may use the rows of 10 to help them calculate the total number of days. ("That's 10, and another 10 is 20, and 21, 22, 23.")

Making Weather Representations After students have had some experience collecting and recording data in the grade 1 curriculum (especially in *Survey Questions and Secret Rules*), they can make their own representation of the weather data. For one month, record the weather data on a piece of chart paper (or directly on your monthly calendar), without organizing it by category. At the end of the month, ask students to total the number of sunny days, rainy days, and so forth, and post this information (perhaps as a tally). Students then make their own representation of the data, using pictures, numbers, words, or a combination of these. Encourage them to use clear categories and show the number of days in each.

VOCABULARY SUPPORT FOR SECOND-LANGUAGE LEARNERS

The following activities will help ensure that this unit is comprehensible to students who are acquiring English as a second language. The suggested approach is based on *The Natural Approach: Language Acquisition in the Classroom* by Stephen D. Krashen and Tracy D. Terrell (Alemany Press, 1983). The intent is for second-language learners to acquire new vocabulary in an active, meaningful context.

Note that *acquiring* a word is different from learning a word. Depending on their level of proficiency, students may be able to comprehend a word upon hearing it during an investigation, without being able to say it. Other students may be able to use the word orally, but not read or write it. The goal is to help students naturally acquire targeted vocabulary at their present level of proficiency.

We suggest using these activities just before the related investigations. The activities can also be led by English-proficient students.

red, blue, green, people, home, house

1. Draw a simple house outline on three sheets of construction paper: one red, one blue, and one green. Draw different numbers of stick people next to each house.
2. Identify the three colors of the houses. Then ask questions about the number of people living in each home.

   How many people live in the green house?
   How many live in the blue house?
   Which house do six people live in?

3. Ask students to match red, blue, and green crayons or markers to the corresponding houses.

coins, cents, penny, nickel, dime, quarter

1. Show a pile of coins. Pick out an example of a penny, a nickel, a dime, and a quarter.
2. Students sort the pile of coins into four groups of like coins.
3. Encourage students to identify the coins by their value.

   Can you show me a coin that is worth 1 cent?
   Can you show me a coin worth 5 cents?
   Which is worth 10 cents, the dime *[point]* **or the quarter** *[point]***?**

imagine, story problem

1. Draw a tree on the board or chart paper. Ask students to close their eyes and picture the tree, and then imagine a bird in the tree. With their eyes shut, ask students to describe their bird by its color. Once students open their eyes, ask them if they see a bird in the tree you have drawn. Explain that the bird they saw was in their imagination; they *imagined* the bird.
2. Draw a swing hanging from the tree. Ask students to imagine a child in the swing and decribe something about the child (for example, boy or girl, what the child is wearing).
3. Ask students to keep their eyes closed as you tell this story:

   Imagine that there is a boy on the swing. He is swinging back and forth. Now imagine that two more children walk up. They are coming to watch the boy swing and to wait for their turn. How many children can you see in your mind now?

4. Ask students to open their eyes and describe the number of children. Write on the board the numbers 1 and 2. Ask them to tell how these numbers relate to the children in the *story problem.*

Blackline Masters

_______________________, 19____

Dear Family,

Our new mathematics unit, *Number Games and Story Problems,* helps build a solid understanding of numbers. As we play number games and solve story problems, the children will explore combinations of 10 (such as 4 + 6, or 2 + 5 + 3), combinations of 20 (such as 3 + 7 + 5 + 5), and combinations of other numbers. They will count and combine objects that come in 2's, like hands, and in 4's, 5's, and 10's. The children will also practice reading, writing, and sequencing numbers up to 100.

Here are some things you can do at home:

- Your child will bring home some of the math games we are playing with number cards, dot cards, coins, and counters. Take time to learn and play these games with your child.
- Look for opportunities to count large groups of objects. You might ask your child to count a handful of pennies, or marbles, or acorns. If several children take handfuls, your child can count each handful and compare them to find which is larger.
- Look for addition and subtraction situations at home. (Numbers under 25 are about right for many first graders.) For example:

  If we have 4 apples, 8 bananas, and 7 plums in the fruit bowl, how many pieces of fruit do we have?

  If you have 20 cents and you spend 15 cents, how much do you have left?

  If we make 4 sandwiches for each person in the family, how many sandwiches is that?

Your child may work out the answers by using counters such as pennies, buttons, or paper clips. Or, your child might draw pictures, count on fingers, write down the steps taken to solve the problem, or work mentally. All these approaches are encouraged in this unit.

Sincerely,

Name Date

Student Sheet 1

Dot Addition

Materials: Dot Addition Cards (20 dot cards)
Dot Addition Board
Sheet of paper

Players: 1–3

Object: Use combinations of dot cards to make the numbers on the board.

Note to Families
You may use any of several Dot Addition Boards that your child may bring home. Although only four cards fit across the board, it's OK to use more to make a number. Just overlap them or let them go off the board.

How to Play

1. Lay out the 20 cards faceup in four rows of five. Place the board beside them.
2. Work together to make each number on the Dot Addition Board with your cards.
3. You can't use a card twice. For example, if you use three 3's to make 9, you can't use all 3's to make 12.
4. You may rearrange your cards at any time until you have made all the numbers on your board.
5. At the end of the game, write the number combinations you made on a sheet of paper.

Variations

a. Play again with the same board. Find a different way to make each number with the cards.

b. Make your own Dot Addition Boards.

Name Date

Student Sheet 2

Dot Addition Board A

9
12
12
15

Name Date

Student Sheet 3

Dot Addition Board B

8
10
10
20

Name Date

Dot Addition Board C

9
10
11
20

Name Date

Student Sheet 5

On and Off

Note to Families

For counters, you might use buttons, pennies, paper clips, or toothpicks. If you do not have a copy of the On and Off game grid, write the numbers in two columns on any paper.

Materials: Counters (8–20)
On and Off game grid
Sheet of paper

Players: 1–3

Object: Toss counters over a sheet of paper. Record how many land on and off the paper.

How to Play

1. Decide how many counters you will toss each time. Write this total number on the game grid.
2. Lay the sheet of paper on a flat surface.
3. Hold the counters in one hand and toss them over the paper.
4. On the game grid, write how many landed on the paper and off the paper.
5. Repeat steps 3 and 4 until you have filled one game grid (eight tosses).

Optional

Your filled game grid shows different ways to break the total number into two parts. Can you find a way that is not shown?

Name Date

Student Sheet 6

On and Off Game Grid

Game 1

Total number: ________

On	Off

Game 2

Total number: ________

On	Off

Name Date

Student Sheet 7

Total of 10

Materials: Deck of Number Cards
(remove the wild cards)

Players: 1, 2, or 3

Object: Find combinations of cards that total 10.

How to Play

1. Lay out 20 cards faceup in four rows of five. Set aside the rest of the deck.
2. Players take turns. On your turn, look for a combination of cards that totals 10. Remove those cards and put them aside. (Put each combination in a separate pile so they don't get mixed up.)
   The 0 card may be included in any combination. The 10 card by itself is one way to make 10.
3. The game is over when no more combinations of 10 can be made.
4. List all the combinations of 10 you made.

Variations

a. Each time you remove cards from the layout, replace them with new cards from the deck.

b. Play with wild cards. A wild card can be any number.

c. Play to make a larger total, such as 12 or 20. Replace each card you use with a new card from the deck.

Name Date

Student Sheet 8

Counters in a Cup

Note to Families

For counters, you can use buttons, pennies, paper clips, beans, or toothpicks. Hide them under any container that you cannot see through. If you do not have a copy of the game grid, write the numbers in two columns on any paper.

Materials: Counters (10–15)
Counters in a Cup game grid
Paper cup

Players: 2

Object: Figure out how many of a set of counters are hidden.

How to Play

1. Decide how many counters to use each time. Write this total number on the game grid.
2. Player A hides a secret number of counters under the cup and leaves the rest out.
3. Player B figures out how many are hidden and says the number. Lift the cup to check.
4. On the game grid, write the number hidden in the cup and the number left out.
5. Players switch roles. Hide a different number of counters. (It's OK to hide the same number of counters more than once in a game.)
6. Repeat steps 2–5 until you have filled the game grid. (Hide the counters eight times.)

Optional

Your filled game grid shows different ways to break the total number into two parts. Can you find a way that is not shown?

Name Date

Student Sheet 9

Counters in a Cup Game Grid

Game 1

Total number: ________

In	Out

Game 2

Total number: ________

In	Out

QUICK IMAGE PICTURES OF 10

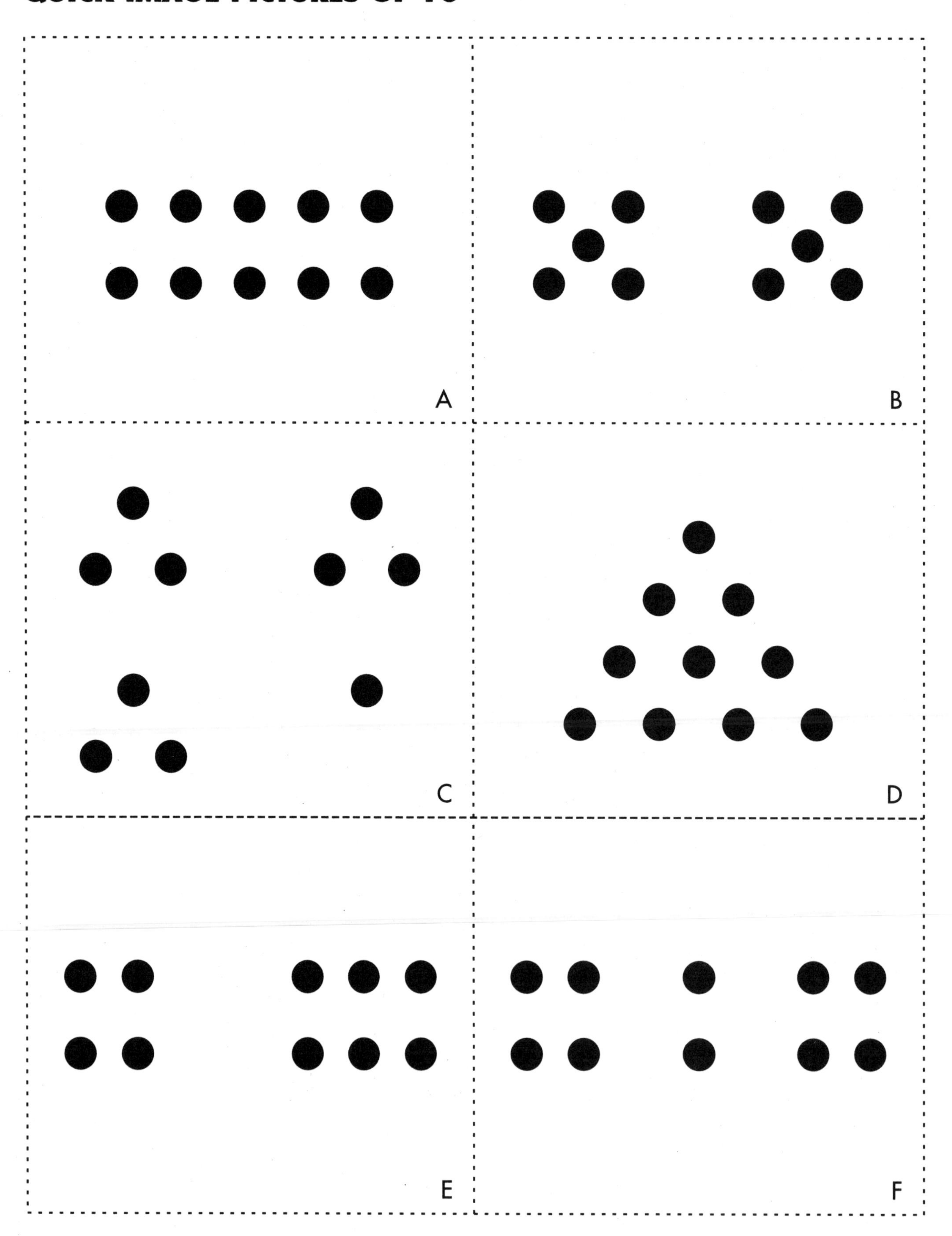

DOT ADDITION CARDS

BLANK DOT ADDITION BOARD

CRAYON PUZZLE 1

I have 6 crayons.
Some are blue and some are red.
I have the same number of each color.
How many of each could I have?

CRAYON PUZZLE 2

I have 7 crayons.
Some are blue and some are red.
I have more blue crayons.
How many of each could I have?

CRAYON PUZZLE 3

I have 8 crayons.
Some are blue and some are red.
I have the same number of each color.
How many of each could I have?

CRAYON PUZZLE 4

I have 9 crayons.
Some are blue and some are red.
I have more red crayons.
How many of each could I have?

CRAYON PUZZLE 5

I have 10 crayons.
Some are blue and some are red.
I have the same number of each color.
How many of each could I have?

CRAYON PUZZLE 6

I have 10 crayons.
Some are blue and some are red.
I have more blue crayons.
How many of each could I have?

CRAYON PUZZLE 7

I have 12 crayons.
Some are blue and some are red.
I have more blue crayons.
How many of each could I have?

CRAYON PUZZLE 8

I have 12 crayons.
Some are blue and some are red.
I have more red crayons.
How many of each could I have?

CRAYON PUZZLE 9

I have 12 crayons.
Some are blue and some are red.
I have the same number of each color.
How many of each could I have?

CRAYON PUZZLE 10

I have 11 crayons.
Some are blue and some are red.
I have more red crayons.
How many of each could I have?

CRAYON PUZZLES CHALLENGE (page 3 of 3)

☆☆☆☆☆ **CRAYON PUZZLE 11**☆

I have 11 crayons.
Some are blue and some are red.
I have one more blue than red.
How many of each could I have?

☆☆☆☆☆ **CRAYON PUZZLE 12**☆

I have 11 crayons.
Some are blue and some are red.
I have three more blue than red.
How many of each could I have?

☆☆☆☆☆ **CRAYON PUZZLE 13**☆

I have 6 crayons.
Some are blue, some are red, and some are green.
I have the same number of each color.
How many of each could I have?

☆☆☆☆☆ **CRAYON PUZZLE 14**☆

I have 8 crayons.
Some are blue, some are red, and some are green.
I have the most blue crayons.
How many of each could I have?

☆☆☆☆☆ **CRAYON PUZZLE 15**☆

I have 14 crayons.
Some are blue and some are red.
I have the same number of each color.
How many of each could I have?

☆☆☆☆☆ **CRAYON PUZZLE 16**☆

I have 15 crayons.
Some are blue and some are red.
I have fewer red crayons.
How many of each could I have?

Name Date

Student Sheet 10

How Many Hands at Home?

Draw a picture of everyone who lives at home with you. Find out how many hands there are.

Show how you solved the problem. Use words, pictures, or numbers.

Name Date

Student Sheet 11

Cats and Paws

There are 4 cats in the yard.
How many paws are there?

Show how you solved the problem.
Use words, pictures, or numbers.

Name Date

Collect 25¢ Together

Note to Families
Your child may play this game with only pennies. Other coins are optional. If you don't have a dot cube (or die), use Number Cards or write the numbers 1–6 on slips of paper, turn them facedown, and draw them from a pool.

Materials: One dot cube
Coins (about 30 pennies, 6 nickels, 4 dimes, and 1 quarter)

Players: 2

Object: With a partner, collect 25¢ in coins.

How to Play

1. To start, one player rolls the dot cube. What number did you roll? Take that amount in coins.
2. Take turns rolling the dot cube. Take that amount in coins and add them to the collection. You may use all pennies, or you may trade coins at any time (for example, 1 nickel for 5 pennies).
3. After each turn, check your total amount. The game ends when you have 25¢.

Variations

a. For each turn, write the number you rolled and the total amount you have so far.
b. Play to collect an exact amount. If the number you roll takes you over 25¢, roll again.
c. Play Collect 15¢ Together or Collect 50¢ Together.
d. Each player has a goal of collecting 25¢.

Name Date

Feet, Fingers, and Legs (page 1 of 2)

Show how you solved each problem.
Use words, pictures, or numbers.

1. There are 7 people in my family.
   How many feet are there?

2. There are 9 children at the bus stop.
   How many feet are there?

Name Date

Feet, Fingers, and Legs (page 2 of 2)

3. There are 2 children in the kitchen.
   How many fingers are there?

4. There are 2 horses and 2 people in the barn.
   Horses have 4 legs. How many legs are there?

Name Date

Student Sheet 14

Coins

Ken and Tara each have 5 cents. Shani and Alex each have 2 cents. How much money do they have in all?

Show how you solved the problem. Use words, pictures, or numbers.

Name Date

Student Sheet 15

100 Chart

1	2	3	4	5	6	7	8	9	10
11	12	13	14	15	16	17	18	19	20
21	22	23	24	25	26	27	28	29	30
31	32	33	34	35	36	37	38	39	40
41	42	43	44	45	46	47	48	49	50
51	52	53	54	55	56	57	58	59	60
61	62	63	64	65	66	67	68	69	70
71	72	73	74	75	76	77	78	79	80
81	82	83	84	85	86	87	88	89	90
91	92	93	94	95	96	97	98	99	100

Name Date

Student Sheet 16

What's Missing? (A)

1		3		5		7		9	
11		13		15		17		19	
	22		24		26		28		30
	32		34		36		38		40
51	52		54	55		57	58		60
	62		64	65		67	68		70
	72			75			78		
81			84			87			90
91		93		95		97		99	

Name Date

Student Sheet 17

What's Missing? (B)

						7			
						17			
						27			
31	32	33	34	35	36	37	38	39	40
						47			
						57			
						67			
						77			
						87			
						97			

Name ______________________________ Date

Student Sheet 18

What's Missing? (C)

						17			
		23							
			34						
				45					
61								69	
							78		
						87			
	92								

QUICK IMAGE SQUARES (page 1 of 3)

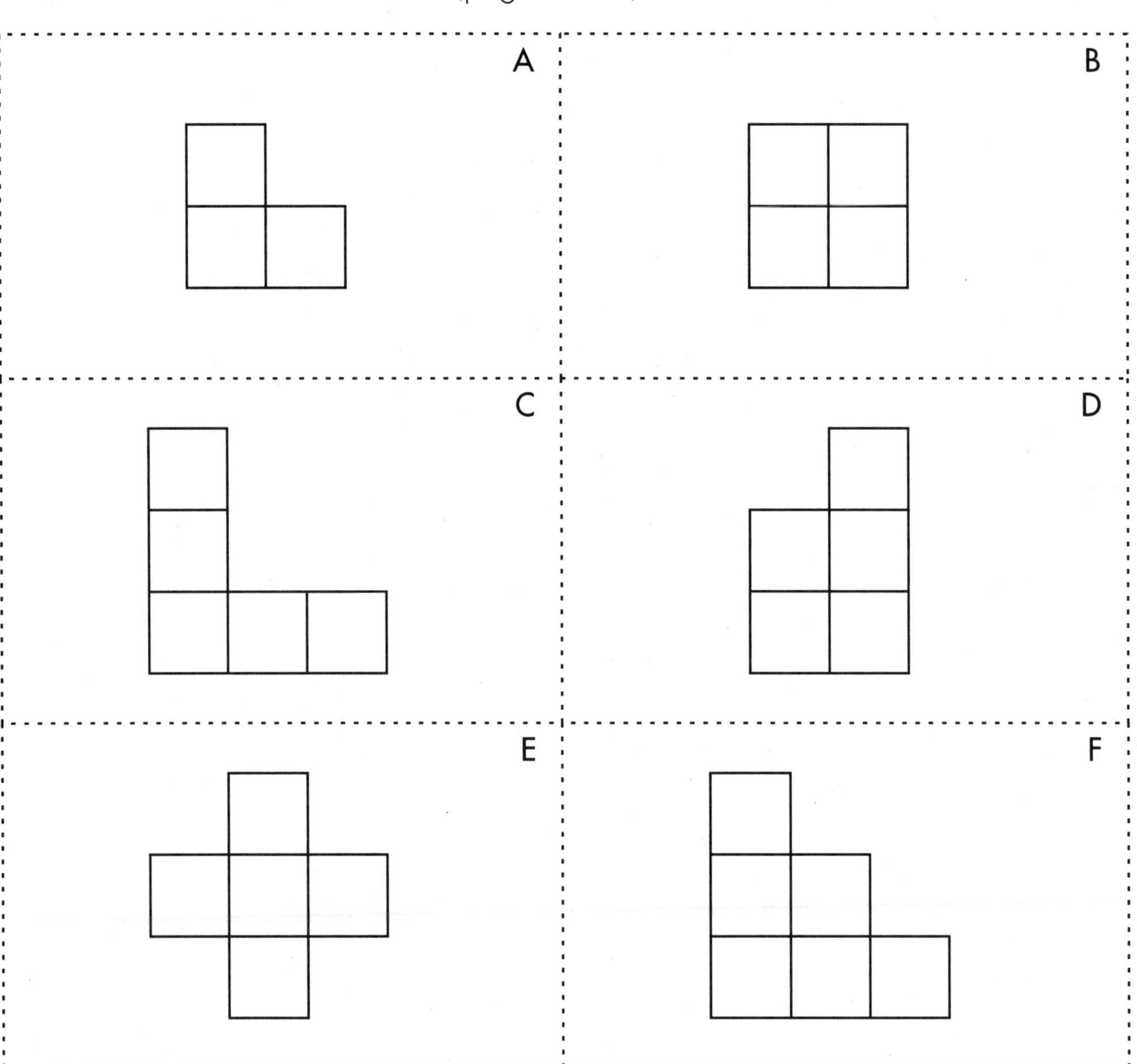

Cut apart the pairs of squares and single squares below to use at the overhead when introducing the activity How Many Squares?

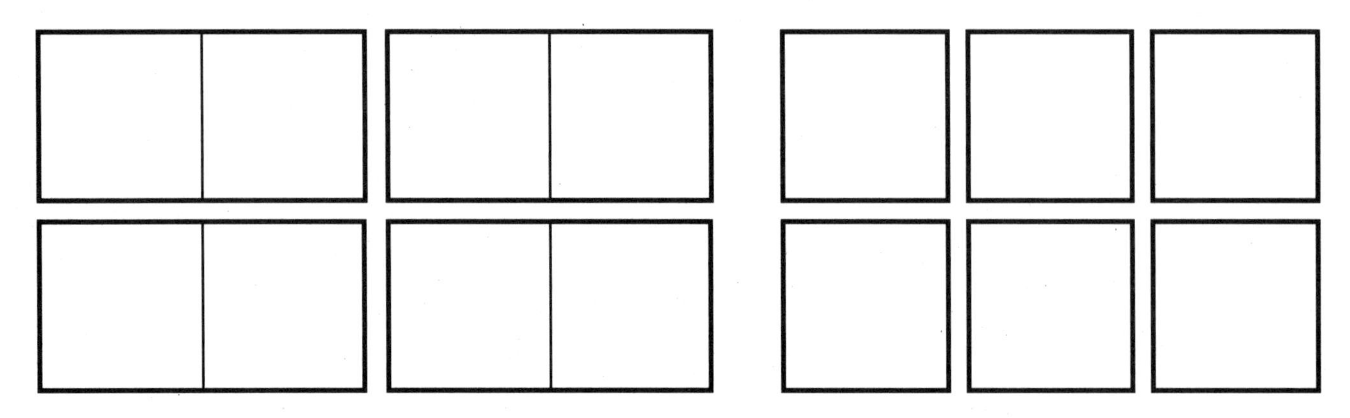

QUICK IMAGE SQUARES (page 2 of 3)

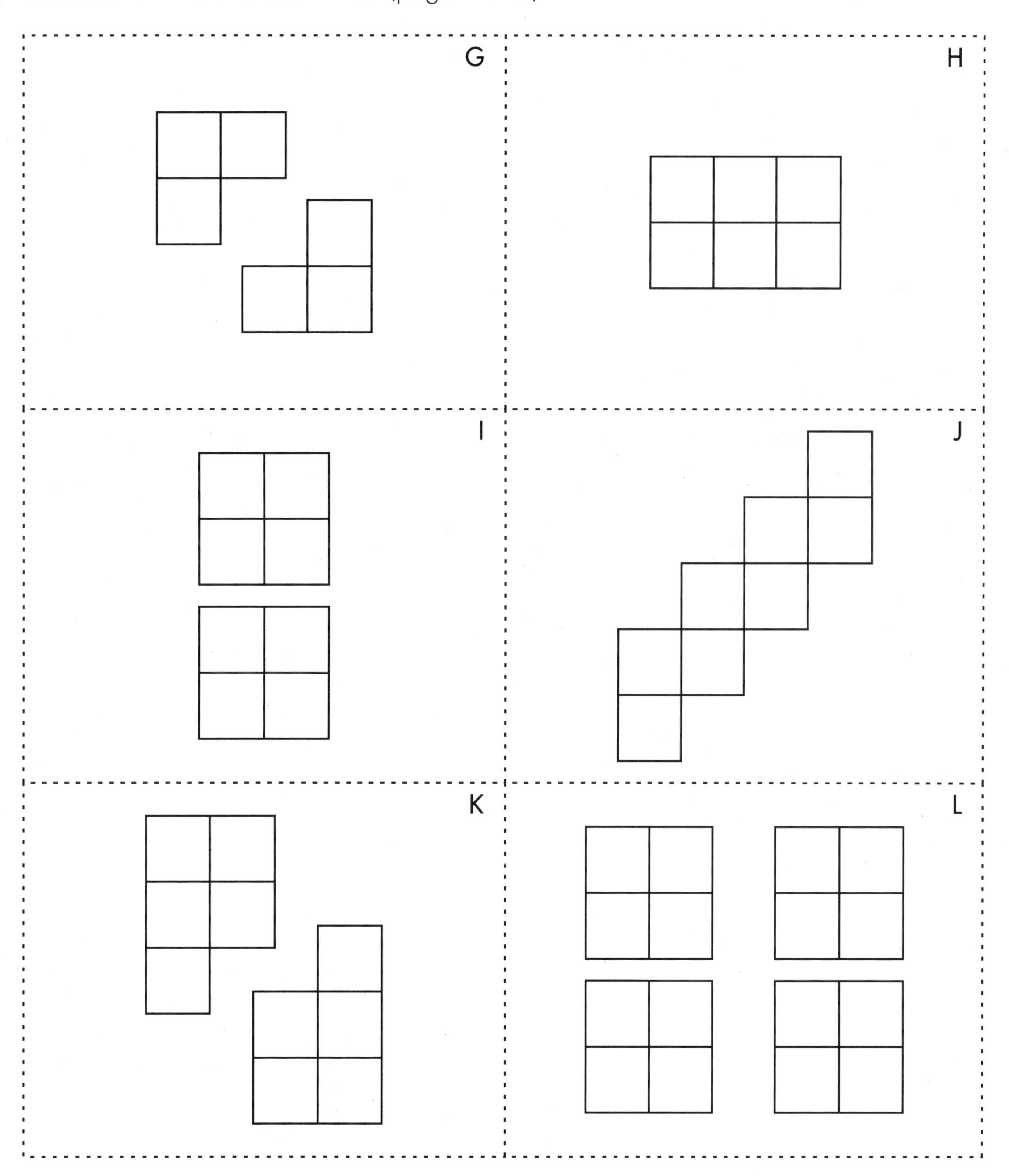

QUICK IMAGE SQUARES (page 3 of 3)

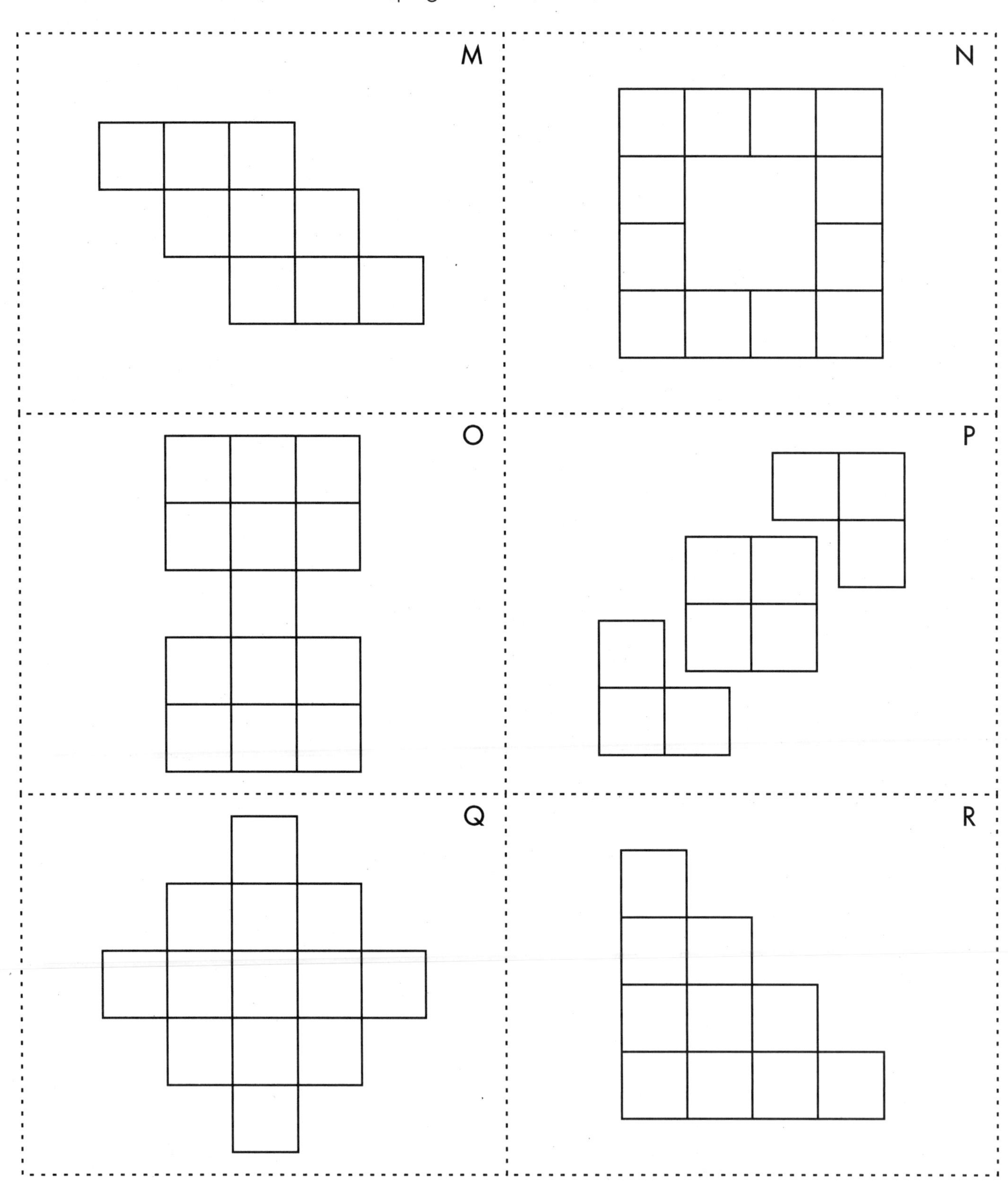

SQUARES (page 1 of 3)

Make two copies of this sheet and cut apart the single squares.

SQUARES (page 2 of 3)

Make three copies of this sheet and cut apart the pairs.

SQUARES (page 3 of 3)

Make three copies of this sheet and cut apart the five-strips.

BLANK 100 CHART

200 CHART

1	2	3	4	5	6	7	8	9	10
11	12	13	14	15	16	17	18	19	20
21	22	23	24	25	26	27	28	29	30
31	32	33	34	35	36	37	38	39	40
41	42	43	44	45	46	47	48	49	50
51	52	53	54	55	56	57	58	59	60
61	62	63	64	65	66	67	68	69	70
71	72	73	74	75	76	77	78	79	80
81	82	83	84	85	86	87	88	89	90
91	92	93	94	95	96	97	98	99	100
101	102	103	104	105	106	107	108	109	110
111	112	113	114	115	116	117	118	119	120
121	122	123	124	125	126	127	128	129	130
131	132	133	134	135	136	137	138	139	140
141	142	143	144	145	146	147	148	149	150
151	152	153	154	155	156	157	158	159	160
161	162	163	164	165	166	167	168	169	170
171	172	173	174	175	176	177	178	179	180
181	182	183	184	185	186	187	188	189	190
191	192	193	194	195	196	197	198	199	200

ROLL TENS GAME MATS

30 MAT

50 MAT

ROLL TENS GAME MATS

100 MAT

Name Date

Student Sheet 19

At the Beach

There were 9 children at the beach. Then 8 children came to the beach.
Now many children are at the beach?

Show how you solved the problem.
Use words, pictures, or numbers.

Name Date

Student Sheet 20

Clay Animals

Mika made 13 clay animals. She gave 7 of them to her friends.
Now how many clay animals does she have?

Show how you solved the problem.
Use words, pictures, or numbers.

Name Date

Student Sheet 21

Five-in-a-Row with Three Cards

Note to Families

Use counters, such as buttons, pennies, or beans, that fit on the squares of the gameboards.

Materials: Gameboard
Counters (about 20)
Number Cards (wild cards removed)

Players: 2

Object: Cover five squares in a row horizontally, vertically, or diagonally.

How to Play

1. Mix the cards and turn the deck facedown. On each turn, turn up the top three cards.
2. Working together, choose a sum that you can make with two of the cards. Place a counter on a square that matches your sum.
   Example: You turn up 3, 1, and 7. You could cover 4 (3 + 1), 8 (7 + 1), or 10 (3 + 7).
   If no move can be made (all possible sums are covered), turn over three more cards. When you use all the cards, shuffle and start again.
3. The game is over when you cover five in a row.

Variations

a. Turn up five cards. Make sums with any two cards.

b. Both players have boards. From the same three cards, you may make different sums and cover different numbers. The game is over when both of you have five in a row.

Name Date

Student Sheet 22

Five-in-a-Row Board A

2	4	6	8	10
10	10	12	12	14
16	18	20	19	17
15	13	11	11	11
9	9	7	5	3

Name Date

Five-in-a-Row Board B

2	3	4	5	6
7	8	9	9	10
10	10	11	11	12
12	12	13	14	15
16	17	18	19	20

Name Date

Student Sheet 24

Five-in-a-Row Board C

20	18	16	14	12
12	10	10	10	8
6	4	2	3	5
7	9	9	11	11
13	15	15	17	19

Name Date

Student Sheet 25

Tens Go Fish

> **Note to Families**
> This game is based on a children's game called Go Fish. Play the game with the Number Cards your child brought home earlier.

Materials: Number Cards (wild cards removed)

Players: 2–4

Object: Find pairs of cards that total 10.

How to Play

1. Deal five cards to each player. Place the rest of the deck facedown.
2. If you have any pairs of cards that total 10, put them down in front of you. Then replace them by drawing cards from the deck.
3. Take turns. On your turn, ask one other player for a card that will make 10 with a card in your hand. For example, if you have a 3, you can ask for a 7.
   If you get what you ask for, put that pair down. Whether or not you make a pair, draw a card. If the card you draw makes a pair, put that pair down and draw again. When you can't make another 10, your turn is over. Any time you use all the cards in your hand, draw two cards.
4. The game is over when there are no more cards.
5. At the end of the game, list the combinations of 10 you made.

Name Date

Student Sheet 26

Write Your Own Story Problem

Choose one of the following.
Write a story problem for it.

8 + 7   4 + 5 + 6   20 + 10

Show how you solved your story problem.
Use words, pictures, or numbers.

STORY PROBLEMS, SET A

Copy one set per student. Cut apart and place copies of each problem in a separate envelope.

SET A	1.	Ken found 12 white shells at the beach. He found 6 brown shells. How many shells did he find?
SET A	2.	Tara had 13 erasers. Then her friends gave her 7 erasers. How many erasers did Tara have?
SET A	3.	The class needs boxes. Tara brought 7 boxes. Ken brought 4 boxes. Then Tara brought 3 more boxes. How many boxes do they have now?
SET A	4.	Two goats had 15 carrots. They ate 8 of them. How many carrots were left?
SET A	5.	Ken had 16 pennies. He lost 10 pennies. How many did he have left?
SET A	6.	Tara had 17 stickers. She gave away 9 of them. How many stickers did she have left?
SET A	7.	I see 4 children and 2 dogs with muddy feet. How many muddy feet do I see?

STORY PROBLEMS, SET B

Copy one set per student. Cut apart and place copies of each problem in a separate envelope.

SET B 1. Shani had 15 beads on a string. She added 10 more beads to the string. Now how many beads does she have?

SET B 2. Shani washed 19 paint brushes. Alex washed 8 brushes. How many brushes in all did they wash?

SET B 3. Alex had 15 pennies in one pocket and 6 pennies in his other pocket. He spent 5 pennies on a sticker. How many pennies did he have left?

SET B 4. Mr. Wing had 14 pumpkins. He sold 11 of them. How many pumpkins were left?

SET B 5. Shani had 25 balloons. She gave away 7 of them. How many balloons did she have left?

SET B 6. Alex made 22 tacos for a party. His friends ate 12 tacos. How many tacos were left?

SET B 7. There are 5 children at one table. Each child has 5 pencils. How many pencils do they have in all?

STORY PROBLEMS, SET C

Copy one set per student. Cut apart and place copies of each problem in a separate envelope.

SET C 1. Tara had 5 crayons. Her father gave her some more. Now she has 9 crayons.
How many crayons did her father give her?

SET C 2. Ken had 8 marbles. Then he found some more.
Now he has 12 marbles.
How many marbles did he find?

SET C 3. 10 people were on a bus. Some more people got on. Now there are 15 people on the bus.
How many people got on the bus?

SET C 4. Tara and Ken collected 16 cans. The next day they collected 14 more cans.
How many cans did they have?

SET C 5. Tara picked 12 apples from one tree.
She picked 13 apples from another tree.
She gave 10 apples to Ken.
Then how many apples did Tara have?

SET C 6. 30 children were playing ball.
Then 11 of them went home.
How many children were still playing ball?

SET C 7. Tara has 2 boxes of 10 markers and 2 boxes of 5 markers.
How many markers does Tara have in all?

STORY PROBLEMS, SET D (CHALLENGES)

These problems are optional. Have available for about half the class. Cut apart and place copies of each problem in a separate envelope.

SET D ☆ 1. There were 13 black birds in a tree. Some red birds flew into the tree. Now there are 20 birds in the tree. How many red birds are in the tree?

SET D ☆ 2. Alex's train has 18 cars. He added some more cars to it. Now the train has 25 cars. How many cars did Alex add?

SET D ☆ 3. Shani gave her rabbits 13 carrots. The rabbits ate some. Now there are 9 carrots. How many carrots did her rabbits eat?

SET D ☆ 4. Shani had 16 pennies in her pocket. She had a hole in her pocket. Some pennies fell out. Now she has 7 pennies. How many pennies fell out?

SET D ☆ 5. The children in Center School have 23 cats and 12 dogs. How many pets do they have?

SET D ☆ 6. Alex baked 30 cookies. His father ate 6 cookies. His brother ate 5. His sister ate 4. How many cookies were left?

SET D ☆ 7. Alex had 42 cents. He lost 15 cents. Shani has 10 cents. Do they have enough money to buy a sticker that costs 30 cents? How do you know?

STORY PROBLEMS, SET E

Copy one set per student. Cut apart and distribute the problems one at a time.

SET E 1. The pencil jar has 8 red pencils, 3 blue pencils, 2 green pencils, and 7 yellow pencils.
How many pencils are in the jar?

SET E 2. Ken had 18 pennies. He spent 7 of them.
How many did he have left?

SET E 3. Tara drew 8 big stars. Then she drew some little stars. Now she has 14 stars on her paper.
How many little stars did she draw?

SET E 4. A frog hopped down 12 steps. Then it hopped down 12 more steps.
How many steps did the frog hop down?

SET E 5. Tara baked 30 muffins. Her friends ate 10 of them.
How many muffins were left?

NUMBER CARDS (page 1 of 4)

0	0	0	0
1	1	1	1
2	2	2	2

3	3	3	3
4	4	4	4
5	5	5	5

NUMBER CARDS (page 3 of 4)

6 	6 	6 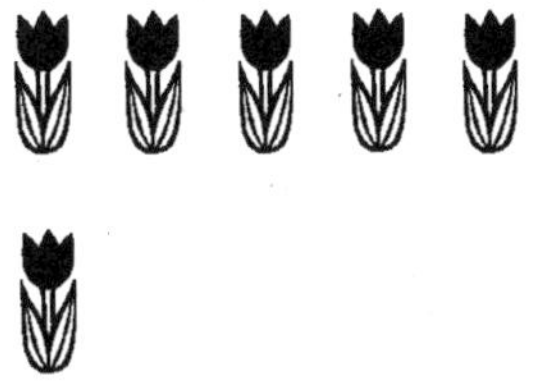	6 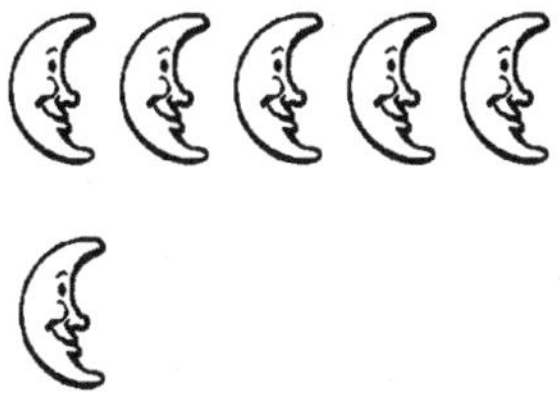
7 	7 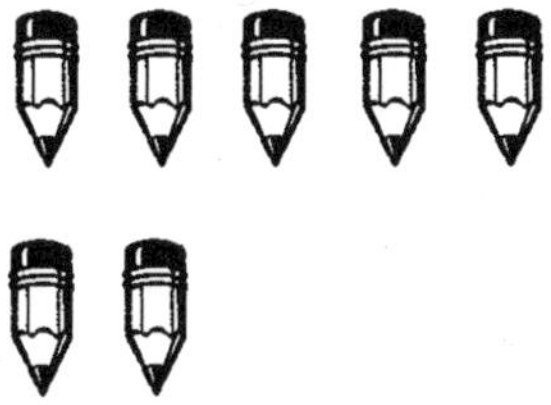	7 	7 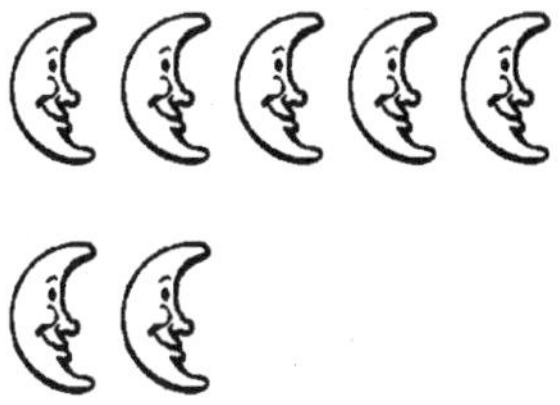
8 	8 	8 	8 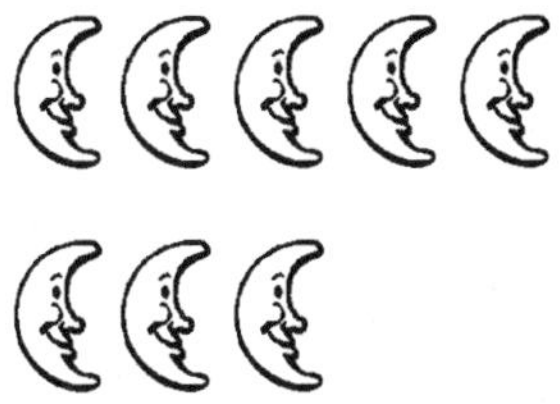

9	9	9	9
10	10	10	10
Wild Card	Wild Card	Wild Card	Wild Card

Practice Pages

This section provides optional homework for teachers who want or need to give more homework than is suggested to accompany the activities in this unit. With the problems included here, students get additional practice in learning about number relationships and solving number problems. Whether or not the *Investigations* unit you are presenting in class focuses on number skills, continued work at home on developing number sense will benefit students. In this unit, practice pages include the following:

How Many of Each? Problems This type of problem is introduced in *Mathematical Thinking at Grade 1,* and students encounter it again in *Building Number Sense* and this unit. Five additional problems are provided here. You can modify any of the numbers and make up new problems in this format, using numbers that are appropriate for your students. Students need not always work with a different total each time they do a How Many of Each? problem. Repeating the same total with different objects gives them more practice with number combinations.

Date

Practice Page A

I have ___15___ blue and gold crayons.

How many of each could I have?

Keep track of your work. You can use pictures, numbers, or words.

Name _______________________________________ Date _______________________

Practice Page B

I have ___18___ circles and squares.

How many of each could I have?

Keep track of your work. You can use pictures, numbers, or words.

Date

Practice Page C

I have ___20___ boats and cars.

How many of each could I have?

Keep track of your work. You can use pictures, numbers, or words.

Name Date

Practice Page D

I have ___20___ squirrels and rabbits.

How many of each could I have?

Keep track of your work. You can use pictures, numbers, or words.

Date ____

Practice Page E

I have __22__ elephants and mice.

How many of each could I have?

Keep track of your work. You can use pictures, numbers, or words.